THE INVENTOR IN YOU

A STEP-BY-STEP GUIDE TO YOUR FIRST INVENTION

CHARLES KANNANKERIL

Copyright © 2018 Charles Kannankeril.
Interior Graphics/Art Credit: Stephen Voss

All rights reserved. No part of this book may be used or reproduced by any means, graphic, electronic, or mechanical, including photocopying, recording, taping or by any information storage retrieval system without the written permission of the author except in the case of brief quotations embodied in critical articles and reviews.

This book is a work of non-fiction. Unless otherwise noted, the author and the publisher make no explicit guarantees as to the accuracy of the information contained in this book and in some cases, names of people and places have been altered to protect their privacy.

Balboa Press books may be ordered through booksellers or by contacting:

Balboa Press
A Division of Hay House
1663 Liberty Drive
Bloomington, IN 47403
www.balboapress.com
1 (877) 407-4847

Because of the dynamic nature of the Internet, any web addresses or links contained in this book may have changed since publication and may no longer be valid. The views expressed in this work are solely those of the author and do not necessarily reflect the views of the publisher, and the publisher hereby disclaims any responsibility for them.

The author of this book does not dispense medical advice or prescribe the use of any technique as a form of treatment for physical, emotional, or medical problems without the advice of a physician, either directly or indirectly. The intent of the author is only to offer information of a general nature to help you in your quest for emotional and spiritual well-being. In the event you use any of the information in this book for yourself, which is your constitutional right, the author and the publisher assume no responsibility for your actions.

Any people depicted in stock imagery provided by Getty Images are models, and such images are being used for illustrative purposes only.
Certain stock imagery © Getty Images.

Print information available on the last page.

ISBN: 978-1-9822-0265-1 (sc)
ISBN: 978-1-9822-0263-7 (hc)
ISBN: 978-1-9822-0264-4 (e)

Library of Congress Control Number: 2018904729

Balboa Press rev. date: 04/25/2018

DEDICATION

I thank my family for being part of this journey.

To my wife, Mary, who has been an inspiration for me throughout this project—I thank you for your support, compliments, suggestions, and especially your patience.

My daughters, Charlene and Crystal, and my sons-in-law, Stephen and Nicholas—I thank you for encouraging, motivating, and helping me throughout this process.

My grandson, Luke, and granddaughter, Lena, my greatest fans—I thank you both for being part of this book and especially for your unconditional love.

My parents, Paul and Treesa—I thank you for teaching me right from wrong and helping me become the person I am today.

My siblings, Joseph, Agnus, Job, Jacob, Elsie, Annie, Mary, Cleetus, James, and Augustine—I thank you for wishing me nothing but the best and for taking pride in what I have accomplished.

It is with great pleasure that I dedicate this book to my entire family.

ABOUT THE AUTHOR

Charles Kannankeril was born in Kerala, India, where he received a bachelor's degree in chemistry. After coming to United States, he received a master's degree in organic chemistry from the University of Massachusetts. Charles was accepted to the doctorate program there, but he decided to pursue engineering to broaden his knowledge base. He received a master's degree in engineering, from the same university in the field of plastics.

Charles has over three years of teaching experience at universities in India and the United States. He has worked in the plastics and rubber industries and has forty-five years of research and development experience in this area. Charles has held various positions in these fields including development engineer, senior development engineer, senior engineering fellow, and director of research and development.

Invention has been a large part of his life. He grew up in India without much more than the essentials. Instead of wishing for the kind of luxury and modern capabilities others had, Charles created his own way by making the best out of what he did have. When he saw or heard about the new technologies and advances outside, he often thought about how to experience and recreate what he was missing. The best way he could do that was by imagining, imitating, and improvising. Charles recognized his hidden talents and skills, nurtured them, and learned how to use them. He learned the necessary techniques to train his mind and prepare it to think and act like an inventor's mind. He set his goals high and started doing things differently with the resources available to him. This attitude of experimentation and determination gave him the creativity, knowledge, and confidence to become an inventor.

During his career, he was granted a total of seventy patents—thirty-seven from the United States and thirty-three international. Currently, Charles has ten patent applications that are pending in the US patent office. Some of his inventions are also protected by trade secrets.

Over the years, he invented and documented over three hundred ideas for new products and processes. His areas of expertise include cushion technology using Bubble Wrap, foam, paper, and on-demand cushion products; food packaging technology related to meat-packaging trays, absorbent pads, and microbial control; medical products such as inflatable hospital beds, ostomy bags, blood-wipe pads, and biohazard material shipping bags; solar-power technology, and polymer formulation and process technology. One of his significant inventions—a high temperature–resistant rubber tape formula that could withstand 3400° F—was not patented but instead kept as a trade secret.

In recognition of his contributions, Charles was inducted into Sealed Air Inventors Hall of Fame in 2004 and received the Engineer of the Year award from the Indian Engineer's Association in 2012.

By sharing his life and research experiences, Charles's goal is to inspire readers to be creative and set their goals high to make meaningful contributions to society by creating a better way of life.

Charlie Kannankeril is one of the most prolific and creative inventors I've met in more than thirty years in business. He has written a wonderful book about a creative young boy who developed a passion for invention and grew up to develop more than seventy patents. It's also a superb guide on how to develop our inventive skills, as told by a top-notch inventor. In his book, Charlie shows us how we all can become more inventive in our professional and personal lives. A worthwhile read by anyone interested in the inventive process.

—William V. Hickey, Retired Chairman and CEO, Sealed Air Corporation

I first met Charlie in the early 1990s after he had already become a legendary inventor at Sealed Air, and I was immediately struck by how humble and curious he was. For someone with his well-earned reputation as a genius inventor he was remarkably approachable; he had time for everyone and was always available to share insights, provide gentle suggestions, and perhaps most important to challenge assumptions.

When Charlie talks about the inventor inside all of us, his philosophy is centered on a trait we all can access but often ignore—the need to be actively curious. Charlie is one of the most curious people I know—always asking questions, then questioning the answers, then using the insights gained to inform his imagination. Charlie would tell you that curiosity feeds imagination, and he certainly has a special gift for helping to unlock the inventor in all of us.
—Ken Chrisman, Division President, Sealed Air Corporation

I believe everyone possesses the potential to be creative and innovative. Some build on another's idea such as by adding a new feature or benefit. Others start with a blank sheet of paper to create a new, game-changing concept.

Having worked with Charlie for many years, I can tell you that he excels in the ability to play both roles. If I were to categorize inventors as novice, apprentice, or master, Charlie is definitely a master. He is known among his peers as an invention machine. Much like Michelangelo describing his sculpting of David where he said he simply chipped away the stone that was not David in order to allow David to emerge, Charlie's imagination of future possibilities brings his inventions to life. Having seen Charlie's imagination and inventive processes firsthand, I can assure you that you are learning from one of the best.
—George D. Wofford, Vice President, Global Core R&D, Sealed Air Corporation

When I think of creativity, innovation, resourcefulness, and imagination no matter the subject matter or obstacle, I immediately think of Charles and his time at Sealed Air. Charles possesses the mind-set and attitude combined with a keen sense of curiosity that allow him to view problems differently and come up with unique solutions. His vast number of patents and trade secrets over his career are a direct indicator that being an inventor is about a drive for continuous learning to solve some of the most challenging issues in industry. Charles's portfolio of work spans a vast array of scientific and engineering disciplines illustrating how to bring out the inventor in all of us from someone who has been living the inventor's life since his childhood.
—Chad Stephens, Vice President, R&D, Product Care Division, Sealed Air Corporation

In the last 25 years that I have known Charlie, I discovered that he was behind some of the game changing inventions in the plastics industry, which transformed the lives of ordinary people around the globe. Innovation and entrepreneurship begin with the identification of problems in our daily lives and finding solutions for them. Charlie began his journey as an inventor by finding solutions to the problems of an under-developed island in India, the land of "Jugaad", by adapting modern technology to enrich lives. His talents eventually flourished in the world's capital of research and development, which gave him a productive environment long before innovation became the engine of growth of the economy. Charlie's life epitomizes the story of how Indian talents, combined with the capital and opportunities of the United States, have created wonders in both countries. Many innovators have told their stories, but Charlie's book, 'The Inventor in You' is more a guide for the next generation of inventors rather than a simple account of his exceptional accomplishments. I am sure that his story will inspire new generations to identify their talents and to develop them to enrich the era of new technology, without losing the human touch
—-T.P. Sreenivasan, Former Ambassador of India, Thiruvananthapuram, Kerala, India.

CONTENTS

Foreword xvii
Introduction xxi

Chapter 1 Good News and Bad News 1
                AIM-IP 3
                      Ambition 3
                      Imagination 4
                      Motivation 4
                      Inspiration 5
                      Persistence 5

Chapter 2 Growing Up with a Different Perspective 12
                Toys 16
                Self-Winding Paper Alligator 18
                Homemade Candle 19
                Learning to Swim 21
                Home Movie without Electricity 22
                Design and Construction of a Manger 24
                School Days 27
                Fantasy 28
                Having Fun 29
                Chemistry Lab 29

Chapter 3	My Experience as an Inventor and the Lessons I Learned...32
	First Experience in the United States 32
	Keep Your Eyes and Ears Open for Opportunities to Invent...34
	Identify the Opportunity—Take the Ball and Run with It 37
	Moving On...41
	It's Not What You Have Learned but How You Use Your Knowledge..43
	Avoid Tunnel Vision and Think Outside the Box........44
	Look at the Big Picture and Find a Creative Way to Invent 45
	Improving Products Using Other Technologies............49
	Invention by Improving a Product for Other Uses.......50
	Invention to Meet Customer Needs.............................50
	Research-Driven Invention 52
	Market-Driven Invention..54
	Invention by Combining Technologies 55
	Having a Vision and Inventing for the Future..............57
	Don't Discount Anything—The Answer May Be Right in Front of You 58
	Helping Other Divisions through Brainstorming..........60
	Getting Ahead of the Competition by Identifying Their Weakness 60
	Solving Customer Problems 62
	Expanding the Business Using Core Competency........63
	Invention from Acquired Technology.........................65
	Inventing through Product Extensions........................67
	Modifying a Product for Other Applications................68
	Adding New Features for Other Applications.............69
	Missing Opportunities by Giving Up Too Early...........70
	Taking Advantage of the Other Features in a Product.......71
	Improvise to Invent 72

One Idea Leads to Another 73
Making Use of Features for Unrelated Applications 73
Creating a Product Is Never the End of an
Invention—It's Only the Beginning 74
Cost Factors in an Invention 75
Invention to Make Life Easier 76
Addressing Customer Requests 77
Special Applications 79
Adapting New Technology 80
Process Inventions 81
About the Company 81

Chapter 4   Invention and Product Development 84
Market-Driven Inventions 84
Research-Driven Inventions 89

Chapter 5   How Do We Invent Things? 91
Why Invent? 91
What to Invent? 91
How to Invent? 92
Chart for Screening Ideas by Matching Features 100
Chart for Screening Ideas for Feasibility 102
Example of an Invention Process 105

Chapter 6   How to Stimulate Your Brain to Prepare You to
Be an Inventor 109
Brain Exercises 109
Using Your Brain as a Database 110
Thinking Outside the Box 112
Triggering an Idea from an Unrelated Event 113
Nonobvious Approaches to Solving Problems 114
Obvious Solutions Made Difficult 114
Taking Advantage of a Failure 115
Challenging the Status Quo 116
Costs 117

	Building on an Existing Invention 117
	Why Didn't I Think of That? 119
Chapter 7	Protecting Your Invention 120
	Prior-Art Search 121
	Documenting the Invention Process 122
	Before Filing a Patent Application 126
	Filing a Patent Application for a Product vs. a Process 127
	Inventorship 128
	Filing a Patent Application 128
	Processing a Patent Application 129
	Granting a Patent 131
	Invalidating a Granted Patent 132
	Other Ways to Protect Your Invention 132
Chapter 8	Analyzing Notable Discoveries and Inventions 134
	Ancient Discoveries and Inventions 135
	Compass 136
	Paper 136
	Gunpowder 137
	Optical Lenses 137
	Printing Press 137
	Electricity 138
	Steam Engine 138
	Telephone 138
	Vaccination 139
	Car 139
	Airplane 140
	Penicillin 140
	Anesthesia 141
	Nuclear Fission 141
	Semiconductor 142
	Computer 142
	World Wide Web 143

Early African-American Inventors Who Changed the World 144
    Thomas L. Jenning 144
    Garrett A. Morgan 144
    George Washington Carver 144
    Jan Ernst Matzeliger 145
    Alexander Miles 145
    Elijah McCoy 145
    Andrew Jackson Beard 145
    Lewis Latimer 145
    George Washington Murray 146
Accidental Discoveries and Inventions 146
X-ray 146
Plastic 147
Saccharin 147
Teflon 148
Velcro 149
Popsicle 149
Matches 150
Viagra 150
Post-it Notes 151
Silly Inventions That Made Millions 152
    Hula Hoop 152
    Wacky Wall Walker 152
    The Top-Down Squeeze Bottle 153
    Pet Rock 153
    Slap Bracelets 153
    Slinky 153
Inventions of the Future 154
    3-D Printing 154
    Robotics 154
    Medicine 155
    Artificial Intelligence 155
    Space Exploration 155

Chapter 9	Roll Up Your Sleeves—It's Your Turn to Invent 156
	Identifying Opportunities to Invent 158
	Home Appliances 159
	Household Items 160
	Preschool and Grade School 163
	Outdoor Applications 164
	Generating Ideas and Solutions 166

FOREWORD

Invention and innovation are in the air these days. Everywhere we look, we see people doing startups, ads for inventor kits, and people taking the plunge into entrepreneurship.

Never in human history has it been possible to have an idea, make it come alive, and get it to market as fast, as cheaply, and as easily as it is today. The amount of technical skill required has plunged at the same that potential market size has exploded.

Tinkerers in small-town India are able to reach global customers; products that didn't exist last year are must-haves for this year's holiday shopping lists; and capital is actively looking for products, services, and ideas to fund.

Also in the air: reinvention. Opportunities to improve on an existing product or approach it from a new angle have opened whole new category lines. I spend hours each month on Kickstarter and Indiegogo seeing what's new, backing projects, and encouraging inventors.

Our home in NYC is filled gadgets and doodads we love. From home-automation products to four Amazon Echo devices, we try to test the latest and greatest. But the gadget I love the most cost us just $10 and has no electronics. It has improved one aspect of our lives in dramatic ways, and every time I use it, I thank the anonymous inventor who made it come alive.

I am talking of course about the hover microwave cover that uses polymer-covered magnets to give the plastic splashguard a place to rest on the ceiling of your microwave while you put your dish in. The result: a clean microwave and one less hassle in your life.

Charlie Kannankeril's career has been all about removing hassles from people's lives. Whether it's a new way of shipping expensive products for multibillion-dollar corporations or helping business owners save a few dollars every day, he's affected the lives of millions of people around the world.

He has more than seventy patents (he had a mere ten when we first met in 1994), working as director of research for Sealed Air, the company that invented Bubble Wrap in 1960.

While all his inventions are impressive, my favorite is the one-handed vegetable bag and dispenser you find in grocery stores. Until Charlie's invention, you needed both hands to pull and tear the bag into which you'd put your veggies. He came along and changed how we shop forever. You'll find the story of the produce bag and dispenser in these pages.

Every time I visit Charlie at his home, he's full of surprises, showing us something new he's working on, something that I know that will help people who likely will never know the man who's helping them.

This book is another Charlie surprise—a surprise and a delight. A surprise because I didn't know he was writing a book or that he was forgoing the typical boring memoir for this kind of invention handbook. A delight because he's converted his decades of experience and deep engineering skills into a practical, useful guide anyone can use.

You will learn step-by-step how to go from idea to market. You will learn how to start thinking like an inventor and solving everyday problems with a non-everyday approach. You will take lots of notes as I did. But most of all, you will learn about five words and Charlie's AIM-IP program: Ambition, Imagination, Motivation, Inspiration, and Persistence.

In my work as a journalist and alleged innovation guru, I've had a chance to speak with some of the greatest innovators of our time: Designer Tory Burch, Microsoft's Bill Gates, Facebook's Mark Zuckerberg, and the cofounders of Instagram—Kevin Systrom and Mike Kreiger. I also worked up close for years with Joe Ricketts, the founder of Ameritrade (who funded a startup I once helped create), and every one of these folks embodies the five words in Charlie's AIM-IP program.

Inventors in small towns in faraway countries may no longer need to come to the US to succeed, but I am so glad that once upon a time, it was essential for a young Charles to make that journey and change our lives in so many ways.

He's done the difficult work of distilling his experience and giving us this invention roadmap. Your picking up the book is the natural next step. Now comes the fun part—the reading—followed by the tougher, yet most satisfying part—the doing. Good luck!

Sree Sreenivasan
New York City
February 2018

(Sree Sreenivasan is the former chief digital officer of New York City, Metropolitan Museum of Art, Columbia University, and social media coach for Columbia Entrepreneurship, Innovation & Design.)

INTRODUCTION

Invention is a technical topic, but reading about it can be fun. In this book, I have shared my experiences in research, my process of inventing, and how this process led me to own seventy patents.

I have presented this material in a simple and easy-to-follow manner and have provided many examples of the process of invention. I wrote this book to help readers recognize their hidden talents and harness their skills to better our collective day-to-day lives.

In this book, I present a systematic, clear approach to the invention process and provide a workshop for potential inventors to achieve their goals. I hope readers will follow the guidelines in this book and invent something they can be proud of.

Invention has been a large part of my life. I was born and brought up on a remote island in India without electricity or any of the modern conveniences. I grew up without much more than the essentials, but instead of wishing that we had the luxury and modern capabilities our neighboring towns had, I created my own luxury by making the best use of what I did have.

When I saw or heard about new technologies and advances outside my island, I often thought about how to experience and recreate what I was missing. The best way I could do that was by imagining, imitating, and improvising. I set my goals high and started doing things differently with the resources available to me. This attitude of experimentation and determination gave me the creativity, knowledge, and confidence to become an inventor.

We do not have to be scientists or geniuses to invent something;

we all have hidden talents that we can harness to become inventors. All we have to do is to recognize these skills, nurture them, and learn how to use them. Inventors can be any age, nationality, or from any walk of life. There is no reason for you to exclude yourself from this group.

People often tell me, "I had an idea once …" They usually come up with many excuses why they didn't or couldn't pursue their ideas that led them to not having seized an opportunity. Later on, they feel regretful, sad, or frustrated when they see someone else creating a product using that same idea. I've also heard people ask, "Why didn't I think of that?" when they see a new invention. The truth is that they could have come up with that same product if they thought like an inventor.

In this book, I share my experiences as an inventor as well as everything I have learned about the process of inventing. You will learn how to think and act like an inventor and learn a simple, effective, step-by-step process for creating and protecting your inventions. This process includes ways to identify an opportunity to solve a problem or improve on an existing idea, develop ideas for possible solutions, work with those ideas to create an invention, and patent your invention.

I have provided helpful techniques I use during the invention process that include illustrations and details to help you understand the process. This book will provide you with useful tools to energize and motivate you to embark on your own invention process.

I want to make the process of invention simple and enjoyable for everyone. I want readers to use this book as a guide in their inventing process while learning how to be creative and utilize their natural skills.

Rather than providing only the technical details of the inventing process, I have blended my life experiences (the good, the bad, and the humorous) in this book so you can learn from my mistakes and successes. I sincerely hope that by using all the tools and going through the workshop in the last chapter, you will achieve your goal of coming up with your own invention.

It has been an absolute pleasure writing this book; I hope it will inspire you and set the stage for your next endeavor. Good luck to you!

GOOD NEWS AND BAD NEWS

CHAPTER

1

Chances are you'll invent something in your life. We all have hidden talents for solving problems, coming up with better solutions, or creating something new. Some of us may be better at these things than others are, but all you need to do to invent something is identify that talent, nurture it, and learn how to use it more effectively. Like any other ability, the more you use it, the better you get at it. And if you don't use it, you can lose it. Consider your talents as tools that can lead to an invention.

You don't have to be a genius or even a scientist to be an inventor. Many ordinary people have come up with extraordinary inventions, and plenty of things are waiting to be created for our everyday lives.

How many times do you hear people say, "I wish there were a better way to do this"? That frequently can give you an idea for an invention. But most of us just say or hear that and do nothing about it, believing we don't have what it takes to solve the problem. We're not talking about designing a rocket ship; in fact, inventing a better product can be as simple as modifying an object you use every day and making it more to your liking. As a consumer, you may know more about the uses of a particular object than the person who designed it does.

Many inventions were the results of making small changes to commonplace tools or items. Not all such design changes are necessarily the work of the manufacturing designers. Many suggestions came from consumers just like you. When you aren't happy with a tool or consumer good, you may come up with an improvement without realizing it. Unfortunately, few people decide to take the next step.

I've heard many people say, "I had a good idea once," but they follow that up with, "but somebody else probably thought of it before," or "It won't work," or "It sounds silly," or "I don't have the time or money to go through with it." The same people may regret not acting when they see their ideas coming to market years later. Just having an idea is not enough; you have to have a strong desire to turn it into an invention. Not all ideas will come to fruition, but you will never know until you try.

Your invention may be an idea for a new product, a new process, or even a new use for something. The idea could be something simple such as folding a piece of paper differently to create a new, unique, or useful shape. It might be doing something differently to make a process easier, faster, or better. Be creative with your ideas. Your invention may be a new toy or a game that brings joy to children, a machine that makes an everyday process easier, a medicine or cure for a disease, a computer program that launches a satellite, something that makes our lives more comfortable, better, and brighter, or even something revolutionary that can change the world.

I mentioned folding paper because I recently heard about someone who had designed a foldable helmet for bicyclists made entirely of paper folded and glued together. I was very impressed by her design and especially with the crush and impact resistance she obtained by using simple pieces of paper in a different way.

You don't have to be highly educated, skilled, experienced, or wealthy to come up with good ideas. Inventors' biggest assets are their sharp minds and their ability to use them. Inventors are always curious about what they see and hear, and they often imagine things differently. Their curiosity frequently leads to questions such as why, why not, and what if. They may look at an object and ask, "Why it is made this way?" They frequently take it a step further: "What if I made it another way?"

When inventors look at things, they don't see just objects; they also wonder how they are made, what else they can be used for, what its current drawbacks are, how it could be improved, and so on. They are not satisfied with the way things are; they are often preoccupied with their goals and sometimes get ideas in the middle of the night or when they least expect them. Inventors are optimistic and do not give up until they succeed.

AIM-IP

I believe inventors have five basic qualities—Ambition, Imagination, Motivation, Inspiration, and Persistence. These are the key components needed for any successful invention process. I put all these components into a formula and called it AIM-IP—AIM for Intellectual Property, the intangible rights protecting the products of human intelligence and creation such as copyrightable works, patented inventions, trademarks, and trade secrets. The goal for all inventors is to secure intellectual property rights for their inventions as that gives them protection and credit for their ideas.

Ambition

We all have ambitions, but how many of us try hard to fulfill them? Sometimes, it takes a personal experience for us to form and focus on an ambition. If you lack ambition—a strong desire to achieve something—and are content with the status quo, you have no reason to do anything differently. An inventor on the other hand has ambition and strives to do things better and challenge the status quo.

When I searched for top inventions by teens on Google, I came across the name of Kylie Simonds of Naugatuck, Connecticut, an eight-year-old cancer survivor. According to the article, while she was undergoing treatment in the hospital, one of the obstacles she endured was the heavy IV bag and stand, which was difficult to push around and whose wire caused her to trip constantly. Based on her experience, her ambition was to find a solution and make patients' lives easier, especially kids receiving chemotherapy or transfusions.

Kylie invented a pediatric IV backpack—a wearable, portable IV machine for kids receiving chemotherapy or transfusions. The bag even comes in colorful designs making it appealing to kids. She calls it the I-Pack. Kylie's design won a prize at the Connecticut Invention Convention in August 2014. She has secured a patent and is trying to raise money to put the backpack into production.

Imagination

Valuing your imagination—the act or power of forming a mental image of something not present to the senses or never before wholly perceived in reality—is another essential quality of an inventor. By imagining things differently, you will be able to ask why not and see things from a new perspective. That will allow you to uncover the hidden features in an object or change it into something with different features. You can visualize things creatively and uniquely, which can be the first step to an invention. Often, working to fulfill a need results in an invention.

Becky Schroeder invented the first glow-in-the-dark writing-and-drawing surface when she was ten. She was doing homework in her family's car while her mom was shopping, and as it got darker, it became difficult for her to see the paper she was working on. "So I thought it would be neat to have my paper light up somehow, and that's when the idea came to me," Becky said.

She started to think of ways to make paper easier to see in the dark. Her imagination eventually led to an invention. She explored various materials and identified phosphorescent paint, which emits light in the dark. She made a paste using this material and painted it on a board. This board provided enough light for her to see and do her homework on a piece of paper placed on top of it. In 1974, Becky Schroeder became the youngest female to be granted a US patent.

Motivation

Motivation—the condition of being eager to act or work—is another key part of this equation; having desire alone is not enough. You need to be motivated to take the ball and run with it. We all want to see our lives made a little easier.

At age nineteen, Blaise Pascal was helping his father, a French tax collector who spent all day conducting complex mechanical calculations. After seeing his father work tirelessly on those calculations, the younger Pascal was motivated to come up with something that would make his father's work easier. He created a wooden box that had sixteen separate dials that when turned would complete addition and subtraction problems quickly and easily. Pascal's invention laid the groundwork for the modern calculator.

Inspiration

Inspiration—the process of being mentally stimulated to do or feel something and especially to do something creative—is perhaps the most important quality in an inventor. With all other qualities in place, this is where they create an idea. Creative minds are always looking for things that can be done differently.

Remya Jose from Kerala, India, was inspired to create a washing machine without using electricity. At age fourteen, she took an old aluminum tub with a cylinder inside to hold the laundry and combined it with an old bicycle to create a pedal-powered washing machine. She invented something that solved a big problem for those without electricity. This environmentally friendly washing machine didn't require electricity and allowed people exercise at the same time.

Remya used her creative mind wisely to combine various technologies and created something unique with multiple purposes and uses. She was not focusing on just one thing; rather, she thought of other benefits while creating her product.

Persistence

Persistence—a firm or obstinate continuance in a course of action in spite of difficulty or opposition—is an equally important quality in an inventor. Failures and disappointments are extremely common during any process of invention. How you react to these events can make or break your momentum. Face setbacks with courage and prepare yourself mentally and physically to move forward. Be willing to keep trying until you succeed.

Dissatisfaction with the status quo coupled with the desire to improve things can also lead to an invention. As a young boy on Hayling Island, off the southern coast of Britain, Peter Chilvers enjoyed a variety of water sports. He became bored with his surfboard and decided to come up with a more exciting design. He experimented with many versions of the surfboard, but he was not pleased with any of them as they did not meet his goal of being different and exciting. But he persisted. One day, he decided to put a sail on his surfboard. In 1958, at age twelve, he created the sailboard.

All the aforementioned qualities need to work in tandem to reach the ultimate goal. A lack of any one of them can make your destination seem so far off.

Just imagine a world without improvements in transportation, communication, and health! Thanks to the people who focused on the qualities outlined in AIM-IP, we all live better lives today. If we thought that we had everything we needed and were content with what we had, we would never see anything change or progress, and we would be stuck in essentially the same world. But that is not going to happen in this fast-moving world in which hundreds of new inventions come out every day. With the way things are changing, fifty years from now, people may look back on our time and be shocked at how we lived without inventions they have now, one or more of which you could be creating.

The good news is that today, inventors have great opportunities to be creative and to invent, especially with rapidly changing technology. Many electronic gadgets we buy these days often become obsolete before some of us can even get used to them. Demand is growing as the result of our insatiable appetite for newer, better, and faster things. More and more technologies are being developed, and this is the best time for anybody to jump into the field of invention.

The bad news is that with such enormous opportunities, everybody is competing to get ahead. There are more than 7.5 billion people in this world, and at least some of them think alike. Therefore, it is very likely that someone else thought up the same idea before you did and can claim the rights to this idea by patenting it. However, it is not always problematic. Since you came up with the same idea, you are in a good position to take that idea to the next level by improving it. Besides, others who came up with this idea may not have done anything with their invention and never tried to protect it by applying for patents. According to current US patent law, a patent is issued to the first person filing for it, not necessarily to the person who invented it. I will discuss the right to an invention and the right to a patent in detail later in this book.

Most inventions are the result of long, hard work. Usually, inventions happen when you are trying to solve a problem or trying to modify a product or a process. Sometimes, they are the result of wanting something different and better. There are many examples of inventions coming about purely by accident due to unexpected results, failed tests, or tests having produced results opposite of those expected.

Many times, a strong desire to have what is missing in one's life or making what one has even better can lead to an invention. Inventors are always asking questions, seeing things differently, wanting things to be better. They remain curious throughout their lives. They avoid tunnel vision and tend to think outside the box. They identify problems rather than ignoring them and look for ways to solve them. A young inventor, M. Tenith Adithyaa, once said, "I do not want to live with problems, I want to solve them." With that attitude combined with ambition, imagination, motivation, inspiration, and persistence, you can be an inventor at any age.

Every time you say, "There must be a better way," you have identified a reason to invent. When you use an item daily, you become an expert on it. Who would know better what might be wrong with it? Most people end their thought process at that stage and don't take it further for various reasons—"I'm not an inventor," "It's not up to me to figure it out," or "People may laugh at my idea." Their lack of confidence makes them give up before they even start. Build up your confidence by realizing you have used the item long enough to know everything about it, including what is lacking and what would make it better. You'll be surprised at what you can come up with when your creativity and confidence unite.

Many famous inventors worked in a certain field and tried to stay in their comfort zones. Others liked to explore any and all areas; their inventions may be less detailed and defined, and they may create more, but their inventions are less specific or specialized. This is similar to medical doctors: specialists versus general practitioners. I have always preferred to be in the second category because I have found more opportunities to invent in it.

It is better for beginners to be in this category. Besides having more opportunities, you can always move from one area to another if you feel you're not making progress. It is much easier to focus on products that we use daily than trying to work on a rocket design (unless you are a rocket designer). Take the challenge in some areas you're comfortable in.

All modern developments were the result of someone having a strong desire to change the status quo and then doing exactly that. Coming up with something new requires long, hard work and the willingness to put up with failures before you get it right. You have to try out every possibility if you want something new, and that takes persistence.

Not all inventions were the result of well-planned and creative research; inventions do happen unexpectedly. Frank Epperson, age eleven, used a stick to mix water and powdered soda on his porch. He went inside and forgot to take the cup with him. Next morning, he discovered that the mixture had frozen around his stirring stick. He pulled both of them out with one tug and found it to be a delicious, cold treat. He decided to sell it around his neighborhood. That was the genesis of the Popsicle.

Many of us probably have experienced something strange happening unexpectedly. That piece of information you just received may be very valuable and could become the basis for your next invention. What you do with that piece of information is important. You can say, "Well, that's strange" and let it go, but you can also be curious about what you can do with that information. You can explore other areas where you

THE INVENTOR IN YOU

can use this information. Once you find a good lead, you can develop this information further to complete your invention.

Some pleasant surprises end up in an invention, and so do some unpleasant surprises. While conducting research, you may get a very bad result or a result entirely the opposite of what you were expecting. Obviously, you are disappointed, and naturally, you will try to redesign your model to get the right results. But true inventors usually react differently; they get curious about the bad results. History shows that many so-called bad results turned into many useful inventions. I will provide some examples of such inventions and discuss them later.

I have purposely given examples of young inventors, ordinary kids who hadn't attended college at the time. They were not scientists, scholars, or geniuses. Just imagine, coming up with these valuable and meaningful inventions at their age having no real experience in life!

Age is not a factor when it comes to inventing things. Sam Houghton became the youngest inventor and a patent holder at age five. Sam's father was using two types of rakes to rake the yard—one to clear up larger leaves and twigs and the other to catch finer debris. Sam tied them together to save Mark the trouble of switching. The result was the Improved Broom, which can be flipped around for different jobs.

When he was ninety-four, Dr. John Goodenough invented and filed a patent application for a new kind of battery that would be cheap, lightweight, and safe. His invention has the potential to revolutionize electric cars and kill off gasoline-powered vehicles. Dr. Goodenough is also the coinventor of the ubiquitous lithium-ion battery; he's still working as a professor in the Cockrell School of Engineering at the University of Texas at Austin and is very active in research.

I was curious to know what the first invention was. My Google search result indicated that it was likely a tool made out of stone that had been chipped or ground down more than 2.6 million years ago; these include arrowheads and spear points.

Though inventors know no age, according to a recent study, 90 percent of all inventors are between thirty-five and sixty-five.

Challenging the status quo; following up on curiosity and using one's imagination; and identifying a problem, a desire, or a need for new or better things all are triggers that can lead to an invention. The result will usually be a unique solution to a problem or the creation of something new and useful.

You'll get excited when you come up with an idea; you'll brag about it to family, friends, and coworkers. You may even think of protecting your idea by applying for a patent, and you'll start dreaming about a financial windfall.

But once you come up with an idea, the first thing you need to do is conduct a prior-art search to see if anybody has claimed your idea before you did. This can normally be done by going through any patent-search websites or even through Google search. In most cases, you will find many matching hits most of which will be related or very similar to yours. And don't be surprised if you find an identical invention. These days, so many inventions are recorded every day that you'll be lucky not to find anything similar to your creation in your prior-art search.

But even if you do identify prior art similar to your invention, you still have an opportunity to invent. An inventive mind will scrutinize all the prior-art inventions for imperfections and look for ways to improve them. You may come up with modifications the inventor missed. Many times, trying to improve an existing invention can lead to a new invention.

When an invention is well protected by patents, it is very difficult to make changes or modifications to improve it. And it is very common to come across matching prior art after you have invented something; you'll feel disappointed when you find out that someone else came up with the idea before you did, but take heart—you can still be proud of what you did and know you came up with the idea independently. That tells you that you have an inventive mind and are in the other inventor's league.

Before you start working on any invention, you need to accept the fact that trying to invent something is not easy. You will have a long and arduous task ahead of you. You will experience many failures on the way, but look at each failure as part of a learning curve. Sometimes, analyzing failures can provide more clues that will lead you to success. Avoid tunnel vision, and think differently, unconventionally, or from a new perspective—think outside the box. Remember the basic formula AIM-IP—AIM for Intellectual Property. Think positive, work hard, don't give up, and try again. You will reach your goal and be successful. You're never too young or too old to start inventing. The great inventor Thomas Edison said, "I have not failed. I've just found 10,000 ways that won't work," "Genius is one percent inspiration and ninety-nine percent perspiration," and "Opportunity is missed by most people because it is dressed in overalls and looks like work."

CHAPTER 2 | GROWING UP WITH A DIFFERENT PERSPECTIVE

My childhood circumstances prompted me to set my goals high. I was born into a middle-class family, but I didn't have much luxury growing up. However, *luxury* was a relative term to me. I was very fortunate to have all the bare necessities, opportunities, and most important, support from my family to think and do things freely and differently. I made the best of everything I had and took advantage of every opportunity I was blessed with. As a result, I had a very busy, fun-filled, adventurous, content, and blessed childhood.

I was born in Kerala in southwestern India. Google will tell you that Kerala is referred to as "God's own country." It is an extraordinary land with almost all elements of nature. Its rich culture and natural scenery make it a traveler's paradise. I grew up in Kumbalanghy, a village on the southern costal area of Kerala. The village is roughly six miles square and surrounded by backwaters on all three sides. Here, we lived a rather primitive life without electricity or any modern conveniences that the neighboring towns enjoyed.

Our village was surrounded by beautiful scenery. The island was full of trees bearing many tropical fruits including coconuts and mangoes, and we heard all kinds of exotic birds chirping during the morning. My family and I lived in a house with a beautiful waterfront view, and we saw sunrises through our bedroom windows. But seeing that every day, I never paid much attention to the magnificence of the place. I never realized how naturally beautiful it was until I moved out and came back years later.

I left home when I was twenty and didn't return for ten years. The first day back, I got up early in the morning to watch the sunrise. I spent hours enjoying the cool breeze and taking in the sights—birds, trees, beach, and carefree fishermen going after the catch of the day. I took several pictures of a sunrise I had seen hundreds of times before. That time, though, it was the most beautiful sunrise I had ever seen.

I enlarged one of those pictures I took in front of my childhood home and framed it. I have received more compliments on that photograph than almost any I have ever taken, and it hangs in the foyer of my house in New Jersey. It allows me to look at that sunrise every day.

I have visited over twenty-five countries, and I have not seen a better or more relaxing place to live than where I grew up. Today, this once-isolated village has been developed into a modern resort area, and it is one the most visited tourist areas in the state of Kerala.

Sunrise View from Kumbalanghy House

Something about this island seems to make people very smart. Even without the modern conveniences back then, this village with a population of 25,000 produced more scientists, doctors, lawyers, engineers, politicians, bishops, priests, and nuns than any neighboring towns—possibly even more than any towns in the state of Kerala.

Kumbalanghy sent more students to the United States and

Europe for higher education than any of our neighboring towns. All my childhood friends achieved notable success. My close friend and classmate from first grade to twelfth grade, Prof. K. V. Thomas, become a member of the Indian parliament and served as a minister for the Indian government; he is one of the highest-ranking politicians in India.

I grew up in a family of eleven—seven boys and four girls. I was the tenth, so you could imagine I did not get much attention at home, but nobody gets all the attention in a big family. I was always very active and full of energy; I always found something to do. Maybe I was trying to get everybody's attention. I guess it worked, because my siblings nicknamed me Thelli-chemmin—"wild shrimp," referring to a type of shrimp that always jumps around in the water.

Attention aside, my parents, Paul and Treesa, took good care of all of us. Ours was a close family. We siblings took care of each other. The older ones paid a great deal of attention to the younger ones. We supported each other and wished nothing but the best for each other. My parents were very loving but also very strict. They always taught us right from wrong and supported us in every way until we could take care of ourselves. Being strict Catholics, my parents insisted we all go to Mass every day.

My father was a strict disciplinarian and made sure we didn't get into any trouble. He was rather overprotective and had a strange way of disciplining and protecting us. One day, my younger brother Augustine came home from school crying; he had been bitten by a dog. My mother tried to console him while dressing his wound. But when my father found out about it, his first reaction was that a dog wouldn't have bitten him for no reason and that he might have provoked the dog. For that reason, my poor brother was grounded for a day.

With that level of strict discipline, we were not allowed to do some of the things that others enjoyed. That helped me cope with not having some of the more luxurious things in life. It also prepared me to handle tough situations, and it gave me the strength to meet challenges.

My parents were very compassionate. One Sunday morning as I was leaving church, I saw a beggar who looked very hungry. My father took my hand and approached the beggar. He gave me some money to give to him. After I gave him the money, my father asked him to come to our house to eat. He was surprised and seemed very happy to be invited.

My mother gave him a home-cooked meal. I can still see the gratitude and the smile on that man's face when he left. One of the most valuable lessons I learned from my parents was the joy of caring and sharing.

The only sources of light at home were kerosene lamps and candles. People hardly traveled at night since there were no streetlights. People used handmade torches—kerosene-soaked rags wrapped around sticks—for emergency travel at night. People carried kerosene to refuel their torches for longer trips. They improvised; they were creative.

The streets were poorly maintained dirt roads, and we had no motorized vehicles in our village. The only modes of transportation were boats, bicycles, and rickshaws. Not having seen any mode of transportation more-advanced than a rickshaw, I thought it was one of the most fascinating objects around.

Once the passengers got into the cart, the operator had to lift the heavy handles; he would end up exhausted and drenched in sweat by the time they reached their destination. But the rickshaw operators never complained; their income was all they had to support their families. By the standard of living at that time, rickshaw rides were expensive, and rickshaws were used only by the rich and by important people such as doctors and priests.

The operators were very competitive; each one tried to make his rickshaw look better than the others to attract more customers. Some operators modified seats to make them more comfortable, or had clear plastic curtains to keep out the rain, or hung kerosene lanterns on them for night trips. Many painted their rickshaws in bright colors. One driver sang loudly to entertain his passengers. Another driver went even further and modified the wheels to make the ride more comfortable. Still another driver came up with a hand brake to help stop his heavy rickshaw, and he shared his idea with other drivers. The rickshaw was invented long ago, but I was very proud of the changes and modifications the operators in our village made to their rickshaws. Today of course, rickshaws have been replaced by cars.

We had candles in case we ever ran out of kerosene, and flashlights were a real luxury—only wealthy people used them. We had a few flashlights in our house, but we didn't use them regularly. I studied and did my homework by the light of a kerosene lantern.

People in neighboring towns used to make fun of us for living without electricity. One might even think that mine was not a suitable environment for an imaginative mind to grow up in, but actually, our village was full of opportunities for any curious and inventive mind. Anywhere we looked, we had plenty of opportunities for improvement and development to catch up with the rest of the world. We were creative; we improvised and tried to make our lives better with what we had. Looking back, we had a good life.

The only daily connection we had to the outside world was the newspaper, which was delivered to the house by a paperboy who went door to door from one end of the village to the other. As soon as the newspaper was delivered, I would rush to read it first. Though there was never much happening in our town, I was curious to find out what was happening in the outside world. While growing up, I did not have many chances to travel to neighboring towns. The first time I remember traveling to a neighboring town was when I was in first grade. I was fascinated by what I saw there for the first time—cars and buses for instance. I wanted to experience how those outside our village lived. From then on, I experimented with ways to imitate or recreate these experiences on my island and tried to improve on them. I looked for every opportunity to make the things around me better. Growing up with this different perspective on life helped foster my creative and ambitious mind, which helped me think, made me resourceful, and taught me to do things imaginatively to get ahead.

Toys

Before I started school, I used to hang around with neighborhood friends. We did not have a toy store in the village, and toys in neighboring town stores were expensive, so we created our own toys. We made whistles from a certain type of tree leaf: it was quite large and was heart shaped. We all learned to make whistles out of them by rolling a leaf in a specific way and blowing into it. When we all played together, it was cacophonous.

One of my friends had a flute his uncle in a neighboring town had given him. My friend was quite good at playing it, and I tried to imitate his notes by rolling my leaves tightly or loosely. I discovered that the number of layers in the rolled leaves and the gap between them was the key to changing the sound. I ended up in making several whistles with different sounds and tried to play along with my friend with the flute. Our cacophony was much less cacophonous after that. My music was no match for the flute, but I came close.

I wanted to have my own flute, but I didn't think I could afford one. I knew it was made from bamboo, and since we had bamboo in our backyard, I decided to try to make one. I cut a bamboo branch and tried to make holes in it similar to my friend's flute. I did not have any tools to do that except a knife, and every time I tried to cut a hole in the bamboo, it split. After several unsuccessful attempts, I examined the holes on my friend's flute to find out how they had been made. My investigation led to an important clue. I saw black char marks around the holes. I was convinced that someone had created the perfectly round holes by burning through the bamboo. So I burned holes through the bamboo. It took many tries as I had to prevent the bamboo from catching on fire. I shaped the burned holes by scraping each one with a knifepoint, and finally, I was a proud owner of a flute.

There was one common toy that everybody in my village knew how to make—a two-wheeled pull cart like a rickshaw. Our village was full of coconut trees that shed baby coconuts all over the ground. These nuts were fairly round and about two inches in diameter. We all looked for a perfect matching pair of these nuts, which we used as wheels for the cart. The leaves on a coconut tree are very long and can grow as long as three feet. Each leaf has a center rib that could be separated from the leafy part to get a long, narrow, and stiff wooden dowel. We used to break this dowel to the right size to become the axle. We would tie mango tree leaves around the dowel and tie the leaf ends together.

The long part of a coconut leaf attached to both sides of the center rib was used as the string to pull the cart. Our toy carts were ready to go. The wheels turned, and the axle rotated inside the leaf. We had a lot of fun running around with these carts. I don't know who the original inventor of this toy was, but we all reinvented them in our own ways. I guess I was hanging around a bunch of talented kids.

We still made those toy carts even after we started school. One day, I thought about improving this toy. I made two similar axles and placed each on both ends of a square frame to create a four-wheel cart. I placed another square frame on the top to shape it like a box and covered the sides and top with mango leaves to make it looks like a car. Pulling this car was more fun than pulling the cart, and my friends copied me and made their own cars. As time went by though, I wanted to do something different with my car.

I had happened to see a toy in a friend's house—a plastic bird attached to the ceiling using a rubber band. My friend's father held the bird and twisted it several times about the rubber band and let it go. The bird would spin until the rubber band fully unwound. That gave me an idea. I attached a rubber band to one of the axles and the frame. By turning the axle several times, the rubber band got wound up. When I let go of it, the wheels started to spin. I repeated the process and put the car on the ground while still holding the wound-up axle. I was surprised to see the car moving forward. Such toys existed elsewhere then, but knowing I could build a toy that moved automatically and independently without any help from others gave me confidence.

Self-Winding Paper Alligator

The next summer, I wanted to play with a new toy, so I began experimenting using the wind-up car design as the base. I made a two-wheeled axle similar to the previous toy and attached it to a rubber band to create a wind-up moving part. After winding up the rubber band by hand, I tied a string to the wound-up axle. When I let the rubber band unwind, the string was wrapped around the axle. Each time I pulled the string to unwrap it, the axle would wind up the rubber band, and the wheels were ready to spin again. That modification allowed me to wind the axle without turning it by hand.

I decided to build another toy using this technique. Using cardboard, I created a cover shaped like an alligator's head and put it over the axle. I threaded the string through a pinhole on the top of the alligator's head so I could pull it from the top. I glued the end of the string to a piece of cardboard to prevent it from escaping inside through the pinhole. I fan-folded a long piece of paper into a V shape to create the alligator's body and tail. I attached the front of the folded paper to the head to create the body. By crushing the fan-folded paper on the back end, I created the tail. I pulled the string all the way to wind up the toy and let it go. The alligator crawled forward several feet until it was completely unwound. And because of my first modification, the string had wound around the axle and the toy was ready to be pulled again. The fan-folded paper on the alligator's body created a wagging effect as it moved. All my friends wanted to play with my toy alligator, and my entire family was very proud of my creation!

Homemade Candle

In our devout Catholic family, we prayed every night before dinner in our prayer room, which had pictures of Jesus and Mother Mary. In front of their images were candles. When I was in third grade, my father gave me the responsibility of lighting the candles before we prayed and extinguishing them at the end of our prayers. I was excited about my new responsibility, and I looked forward to it every night. I wanted to impress my parents by doing a good job.

I remembered seeing how the flames on tall candles were put out in our church. They used an inverted cone attached to a long stick. Placing the inverted funnel over the top of the burning candle put out the flame. So I attached a funnel we used to pour liquids into bottles to a long stick I got from our backyard. I wanted to surprise everybody with this gadget, so that night, I sat behind everybody during the prayer.

After the prayer, without moving from my seat, I extended the stick and snuffed out the candles with the funnel. It was a big hit that night, and I got many compliments from my parents and siblings. They were all impressed by what I had done as a third grader.

The metal plates we set the candles in were full of melted wax. My father asked me if could clean those plates, so I took them outside and started to scrape the wax off the plates. The wax of course was hard and very difficult to scrape off. After trying for some time, I decided to look for something sharper to remove the wax with, and I left the plates in the sun. I came back and noticed that the wax had turned soft and I could remove it from the metal plate rather easily. I gathered all the wax and shaped it into a ball. I wondered if I could make more-complex shapes if I left it in the sun for even longer. I discovered I could, and I molded it into more-complex shapes.

One day, I put the wax on a metal plate and put a kerosene lamp underneath it. After the wax melted, I poured it into different containers to create those shapes when the wax solidified. Then I thought of finding a long cylinder with which I could craft new candles; all I needed to figure out was how to get a wick inside the mold.

Well, bamboo is hollow—a perfect cylinder. I got a length of bamboo from the backyard and made a cylinder with open ends. I put a wick through a piece of cardboard and threaded it up the bamboo. I tied the other end of the wick to a narrow bar placed across the top of the cylinder. My mold was ready. I melted the wax and poured it carefully into the cylinder. I had to do that in several steps. The wax took a long time to cool and solidify in the cylinder. When it became semi-solid, I dipped the cylinder in a bucket of water to cool it faster. Once it solidified, I untied and removed the top bar and pushed the wax from the top. It came out as a perfect candle. It was a lot of work, but I was very proud of my first homemade candle. That night during our family prayer, I lit only one candle—a very special one I had made by hand.

Learning to Swim

As I was growing up, my older brothers and their friends swam in the lake in front of our house. They were playing games while swimming and having a lot of fun in the deep water. I wanted to join them so badly, but I didn't know how to swim. I tried to imitate them and float on the water but without any luck. I tried to hang onto a piece of wood floating nearby, but my weight dragged it down. I wanted to float on the water so badly that I tried larger pieces of wood. I floated better each time I found a bigger piece of wood. Finally, I found one that was bigger than me, but I somehow managed to drag it to the water and floated as I held onto it. But it was too big for me to freely move around in the water and join in the play. I was determined to join the group and play in the water. With that motivation, I started to look around for small objects that floated and could also handle my weight. We had never saw or heard of floatation devices in those days.

During my search, I noticed we had several aluminum pots we used for storing water. They were gallon-sized and had a narrow opening on the top. They floated freely when empty. When turned upside down and placed in the water, a small pot could support my body's weight. I even asked one of my friends to hold onto the pot, and we both were able to float holding onto a single pot. At that time, I did not know how that small pot was able to help float that much weight in the water.

Later, I learned that all the air trapped in the pot was more than enough to counteract the body weight of two people. The pot's narrow mouth helped keep the air inside when the pot was upside down in the water. I found out that this was the best option I had to help me learn to swim and play in the water with my brothers and their friends. With the pot, I was able to float and move around freely. But the pot's surface was very smooth, so it was difficult to hold onto the pot. I kept slipping away from the pot. And because it was expensive, I didn't want to damage it while trying to create a handle.

So I tied a rope around the pot and made multiple loops that would serve as handles for more than one person. I also made a pair of handles that were loose so I could slide my arm through and could swim without holding onto the pot. From then on, I went into the deep water and played with the older kids. I'm sure that many people used similar floatation devices in the past, but I am glad I did it systematically using only my knowledge, determination, and experimentation skills and without any help.

Soon, I learned how to swim without the use of these pots, and I became a very strong swimmer after a couple of years. One day, my friend challenged me to swim across the lake, touch the other side, and come back. The lake was about a mile wide, and I knew it was risky swimming both ways in one stretch. I knew my parents and older siblings would get mad at me if I even tried something like that. But I decided to take the challenge, and I practiced swimming for several weeks. One day, when nobody was home, I started my journey. I knew it was very dangerous, and I was very nervous. As a precaution, I decided to take my aluminum pot flotation device with me; I knew that would slow me down, but I opted to be safe rather than sorry.

My journey started before noon, and I got to the other side in a couple of hours. I sat down and rested. I was very tired, and it took me longer to get back. It was very late when I finally got back home. Luckily, nobody was there to receive me after a successful but dangerous swim. I felt very proud knowing that I accomplished something great that day. For the fear of punishment, I never shared this incident with my parents, although some of my siblings learned about it years later.

Home Movie without Electricity

As I mentioned, our village didn't have electricity, so I didn't know what I was missing. The closest experience I had had with electricity was my family's flashlight, which I played with more than anybody else in our family did.

One night, I turned the flashlight on and put it in my mouth; I saw my face turn red in the dark. The next night, I walked around the neighborhood with this flashlight in my mouth. I managed to scare a few people and especially the little kids, who screamed with fright. I

THE INVENTOR IN YOU

felt guilty and immediately turned the light off and identified myself. The kids felt much better after seeing that it was me.

We had to go to a neighboring town with a movie theater to watch movies. One of my great uncles was an entrepreneur. At that time, he was the only one in our village with a bachelor's degree, a BA; we proudly called him Uncle BA. He wanted to build a movie theater in our village, so he bought an old kerosene-powered generator. After working on it for months, he was able to power a movie projector.

He built a movie theater near his house; it was probably the most primitive movie theater ever, but my uncle was creative. The walls and roof were made of woven coconut leaves. He had three sections of seating with three different admission prices. He made sure that everybody regardless of wealth would have a chance to see a movie while he made a profit. The poorer people were admitted for a very low price—a family of five was admitted for less than two cents, but they had to sit in the front close to the screen. They sat on the floor, which was sand; it was like sitting on the beach.

Behind them was a higher-priced section for middle-class people who sat on benches without back supports. The rich people got the best seats, which were in the back, with comfortable cushioned chairs, but they had to pay more than ten times what those in the front section paid. My uncle made many people—rich and poor alike—happy and at the same time made a lot of money by using his creative mind.

I saw my first movie at his theater when I was in fifth grade. I was curious to learn how the movie process worked. I saw where and how the movie was projected on the screen; the theater used a large spool of movie film, which was wound to an empty spool at a certain speed, and the movie was projected onto the screen by placing a powerful light source on the film while it was moving. Sometimes, the film broke during the show, and they fixed it by splicing the pieces together after cutting off the broken end.

I had an idea how the movie worked. One day after watching a movie, I found a few discarded pieces of broken film. The next day, I conducted a test. I closed all the windows in our room to make it as dark as possible. I pointed the flashlight at the wall and put the film in front of the flashlight. The image was fuzzy, but I learned to sharpen the image by moving the film away from the flashlight. The piece of film I had was about a foot long. I took two narrow pieces of cardboard and placed one on the top of the other. I cut a square hole in the middle of both to match the size of each frame. I inserted the film between the boards and pushed it through until the first frame appeared in the square hole. My first test was repeated until I got a clear image on the wall. The film was pushed slowly, and I was thrilled to see a moving image on the wall. The next time I went to the movie theater, I asked the projectionist if he had longer pieces of broken film, and he gave me several feet of it. With that, I continued to improve my home theater.

I wanted a brighter image, so using a mirror, I created a bright source of light by reflecting outside sunlight and focusing this light beam toward the window. I placed cardboard around the whole window and cut a hole to allow the light to enter. The rest of the room was dark. I had replaced the flashlight with this brighter light source all without electricity. That day, I entertained my whole family and some friends with a home movie!

Design and Construction of a Manger

When I was young, I always looked forward to making an outdoor manger during Christmas time. Most Christian families in my neighborhood built mangers in their homes, but I wanted something different and better, so instead of making only a manger, I designed a whole town with the manger in the center. I did not know much about where Jesus was born, so I decided to build a unique town using my imagination.

Each year, I started my construction about two months before Christmas. The manger platform was about forty square feet and was about four feet from the ground. The platform was filled with wet sand about six inches deep. My design called for an imaginary town with a rice paddy, modern streets with electric streetlights, playgrounds with water fountains, and in the center, a small manger with cattle around where the statue of baby Jesus would be. I knew these modern features weren't available during Jesus's time, but I was trying to create things we didn't have either; I wanted to see what it would be like living in a modern world.

With help from my father, I calculated the time needed for rice to grow to the right height and ready for Christmas Day, which was approximately a month away. I soaked the seeds in water for two days before spreading them on the rice paddy area. In about three weeks, the fields in this small town were covered with about four-inch tall rice bushes. It was such a beautiful sight to look at the miniature paddy fields built as part of the town design.

Two other attractive features of my design were the streets and streetlights. All our village streets were dirt roads, and I wanted to make good-looking roads. After laying out streets of wet sand, I covered them with a paste concoction of rice flour, water, and a few drops of black ink from a fountain pen!

I had to work very hard on the lighting. My father gave me money to buy some flashlight batteries, tiny flashlight bulbs, and some wire. My light poles were sticks with crossbars. I put a lightbulb on top of each pole, and I connected all the bulbs with the wire that I connected to the batteries hidden behind the crib. I used a rubber band to attach the wire to the battery on both ends. The removable rubber band on one side was used as a switch to turn the circuit on and off.

The highlight of the manger was a miniature park and a water fountain in the middle. Without any electricity, I could not pump water through the fountain's nozzle, but I relied on gravity for that. I tied a bucket to the top branch of a tree next the manger and filled it with water. I placed one end of a rubber tube in the bucket and attached the other to the bottom of an old ballpoint pen to serve as the fountain's nozzle. The hole in the end allowed too much water to flow through and did not give off the effect of water flowing in a typical fountain. I couldn't think of any other parts that would do the job, so my only choice was to somehow modify that piece of plastic. Without modern and proper tools, it was difficult working on something like that. But after thinking about it for a few days, I decided to close the top hole by heating it over a candle, and then I made a much smaller hole by pushing a needle through the molten plastic.

The water flow started through the rubber tube by siphoning it from the bucket. I was able to control and even stop the flow by pinching the rubber tube between two pieces of metal held together by a tight rubber band. This mechanism became the on/off switch for the water. The bottom end of the rubber tube was connected to the newly made nozzle. I connected the nozzle to the rubber band and then buried it under the sand, leaving the nozzle tip exposed. When the switch was turned on, the fountain shot a thin stream of water about a foot high. I was able to operate the fountain continuously for about two hours with just one bucket of water.

I was always thinking about ways to improve my design, so every year, I made a few modifications to my design. Some of the modifications included using colored water in the fountain, hiding a lightbulb under the water to shine on the water flow at night, and giving the lightbulbs different colors by wrapping them with semitransparent colored paper.

One change I was particularly proud of was to redirect the water from the fountain to the paddy field instead of letting it go to waste. I collected all the water from the fountain in a reservoir at the bottom, and part of the water irrigated the paddy field through a tube. I guess I created a miniature underground water sprinkler system. I did not do any of this to save the environment; those days, nobody was concerned about the environment. I did that because it was more convenient and less work for me to maintain the paddy that way, but I was glad not to have wasted water.

My friends and neighbors would hang around the manger and watch the streetlights and the water fountain. I got so many compliments from my family and the people in the neighborhood for building that colorful, innovative, and active manger. Some people even tried to imitate my work by building their own versions. I felt a lot of satisfaction after seeing all these positive reactions from others. Though there were no rice paddies, streetlights, or fountains where Jesus was born, I remember hearing people say that my imaginary town was unique and looked beautiful. I am glad I could entertain others at that young age.

School Days

I attended our village school, which offered classes from first to tenth grade. It was one of the toughest schools imaginable. The teachers were very strict, and the students were punished in every way imaginable for being late, not paying attention in class, not completing homework, providing wrong answers, or any mischievous behavior. The punishments were physical and verbal; sometimes, we were humiliated depending on what we did wrong.

I remember having severe anxiety attacks and a racing heart every day on the way to school. Any time we were late, we would be sent to the headmaster's office. He would make us kneel on sharp gravel spread on a concrete surface for fifteen minutes or more depending on how late we were. I got that punishment many times.

But one of my teachers took the concept of punishment a unique step beyond. He thought and did things differently than other teachers. Once he started his class, he wanted to conduct the class without interruption. He argued that all his students would make some mistakes during his class, so instead of punishing each student right in the middle of teaching, he asked all students to stand up, and he beat everyone with his bamboo stick before class even started. The only fair thing about this was that the number of beatings we received depended on how good or bad we were as students, and he always had a good idea about that. I was very lucky to have received minimum beatings. In those days, I thought it was normal to punish students the way they did in our school.

In spite of this boot-camp education, I am proud to say that this school produced more highly qualified people than did any school in our neighboring towns. Recently, I went to our fiftieth high school reunion and was very pleased to see many of my classmates in very successful positions.

My education there gave me the strength to handle any situations and the confidence to face any challenges.

Fantasy

Our only source of news was the newspaper and newsreels played in theaters before each movie. I did not get to see movies very often. The few times I went to the theater, I paid attention to the newsreel, which showed famous people such as presidents, prime ministers, and movie stars getting out of airplanes and being received with applause from crowds. These famous people were always well dressed mostly in suits and wearing sunglasses. I wanted to experience what they did—getting out of an airplane and waving to people. I always set my goals high. I wanted to fly even though at that time, I had never been in a car.

I wore shorts and a short-sleeved shirt to school, never long pants or long-sleeved shirts, and I had never seen a suit or a tie in real life. One summer, my older brother, Lieutenant K. P. Job, came home for vacation. He was in the Indian navy and used to come home for vacation every couple of years. Being in the navy, he got to travel to many countries, and he was familiar with the advances in technology and conveniences of the modern world. He used to share all his experiences of visiting other countries, and we all sat around for hours listening to his stories.

When he visited one summer, he had packed a suit and tie in his luggage that he wore for his formal meetings and parties at work. One day when nobody was looking, I tried on his suit and tie and put on his sunglasses. Of course, his clothes were too big for me, but I folded the sleeves to fit me as best I could. I stared in the mirror and pretended I was getting out of an airplane smiling and waving. My fantasy continued for several minutes. I thought I was somebody important, and I got a taste of how an important person looked and felt. For a moment, I thought I was a celebrity. I wanted that so badly, and I was very happy to see myself looking like a famous person. That experience did something to me—I became determined to be somebody famous when I grew up, so I set my goals even higher.

THE INVENTOR IN YOU

Having Fun

I hung around with a few mischievous friends in high school. We liked picking fruit from a tall mango tree in the schoolyard. The tree belonged to our church, so we were not supposed to do that. But our gang usually waited until all the teachers had gone home to get some of the fruit. But it was difficult to climb the tree and pick mangoes since the tree was very tall and the mangos were at the ends of each branch.

One of the boys had a slingshot he tried to shoot down mangos with. Since he was not successful, everybody got a chance, but we did not get any mangos even after many tries. I thought about using something other than the small stones we had tried. I shot a stick out of the slingshot at a bunch of mangoes a number of times, and one time, some mangoes fell. All my friends wanted to try my stick trick, and we ended up with more mangos than we could eat. I realize now that I had converted the slingshot into a compact bow and arrow.

But I felt guilty for picking the forbidden mangos. I used to go to church every day and went to confession every week. The next week, I told the priest about my new sin of stealing mangoes from church property. The priest happened to be the manager of all our church properties including our school. After hearing about the mangos, the priest got angry; he threatened to tell my father what I had done. I was so scared of what my punishment would have been that I never shot another mango. But I wonder now what happened to the confidentiality of the confession process—the seal of the confessional.

Chemistry Lab

After graduating from high school, I entered college at age fifteen. Chemistry was my favorite subject; I used to do chemistry experiments for fun. One day, I managed to get some chemicals from the chemistry lab that changed color in the presence of sugar. I tested this formula by introducing a trace of sugar water. I thought I could use this technique to diagnose diabetes in patients. I knew that one of my distant relatives was thought to be borderline diabetic. I checked with my parents to see if it was okay for me to ask for a urine sample for my test. My parents got upset and asked me to stop what I was doing. They had good reason for reacting that way; they knew that I wasn't qualified to conduct such medical tests on humans and that any false results could lead to unnecessary worries. I knew I was wrong to attempt such tests, so I never tried those types of tests again.

One of the tests we had to perform during the chemistry final examination was to identify two unknown chemicals. To complete the test, we had to make a byproduct of these two chemicals, name it, and present it to the examiner. My college's lab was not well maintained, and the chemical supply was old and inadequate. The procedures we used to identify the chemicals worked when the chemicals were in good condition, and in my case, one of the chemicals I had to test was of good quality, so I identified it easily, but the other was old and bad, so the standard procedures I used did not give me any clue of what it was.

After struggling for a while and using all my chemistry knowledge, I came up with two possibilities for the chemical but could not eliminate one or the other. I went over all the properties of the two possible chemicals to find out how they differed. After struggling for a while, I remembered one distinctive difference between them, and I realized there was a simple test to identify one. I conducted that test, and the result was negative. By the process of elimination, I was able to identify it.

So I had identified both chemicals, but I still had to make a byproduct by mixing the chemicals. Knowing that the second one was of bad quality, I felt sure I would not get the result I needed, and that indeed happened. I kept all the test samples as proof. In the report, I documented all the tests that I had conducted, all the steps I had taken to identify the chemicals, and the reasons for the additional tests. I also documented the reason for identifying the bad chemical by a process of elimination and the reason for not producing the byproduct.

I thought I would fail the examination for not producing the byproduct. When I got the exam results back, I saw the grade on the front page with an A+ and the comment "Great Work and Excellent

Execution." I hadn't expected that; I was very excited to see that comment from my professor. Apparently, he liked my reasoning and the way I had come to a systematic conclusion. That comment gave me great confidence. The lesson I learned from that was that how you do certain things is just as important as what you get out of doing them.

Rarely do you find a straight path to a solution to your problems. In most cases, you will hit a roadblock and not know how to proceed. You will fail no matter how many times you try to move along that same path because it's the wrong path, and tunnel vision can blind you to that. That's when you have to think outside the box, to explore other paths even if they are longer and more complicated.

At age nineteen, I received a bachelor's degree in chemistry from Sacred Heart College in Kerala, India. I was offered a position as a junior lecturer there mainly teaching students in the lab. After working there for a year, I made the most important decision of my life—to go to the United States for higher studies.

CHAPTER 3

MY EXPERIENCE AS AN INVENTOR AND THE LESSONS I LEARNED

I have always looked forward to taking on challenges and enjoying the outcomes; invention is a big part of my life. This chapter is about my industry experience and how it helped me become productive and successful. I have also included some familiar inventions conceived by the people with whom I worked. I will analyze these inventions and discuss their origin and reasons for their invention, the approaches made on the path to creating them, the results, and the lessons learned.

I have also provided pictures of some of the prototype samples I created for these inventions. Please note that due to the proprietary nature of this information, I had to limit the discussion to only certain inventions. The inventions I discuss are commercialized, patented, or otherwise published. By sharing these experiences, I hope to inspire many and especially those of the younger generation who always wanted to invent but have never taken any initiative or had the opportunity.

First Experience in the United States

I came to the United States in 1967. I was offered a schlorship from St. Anselm's College in Manchester, New Hampshire, which covered my tution and board. Those days, it was rare for people to go abroad for higher education. The Indian government allowed you to carry only $8 out of the country, so you would have to find other means to support your travel and other expenses.

I had the choice to come to the United States by ship or air. The trip by ship would have taken about a month; one of my cousins had done that. His father had given him a large bunch of plantains to eat onboard. A plantain is a large banana common to India, and it is very filling. The bunch his father had given him had thirty plantains, and he ate one a day: he finished the last one on the day his ship reached the United States.

I traveled by air, my first flight. I was not a bit nervous; all I felt was excitement. My college was in Manchester, New Hampshire, and getting there required five flights. I flew from Kochi to Mumbai and then to Frankfurt, Germany. I had an overnight stay and dinner in Frankfurt at the airline's expense. During dinner, the waiter asked me if I wanted anything to drink. He showed me a list of drinks; the only one I recognized was beer. I had never had a beer, so I ordered one. After dinner, I gave my dinner coupon to the waiter, and he gave me a bill for $2. It turned out that my coupon did not cover any alcoholic drinks. I was shocked; I had only $8, and that was supposed to last until I got settled in the new country. I learned my lesson; I was determined not to eat or drink until I reached my destination.

My flight to New York was delayed, so I missed my flight to Boston. It was late, and I ended up taking the last flight to Boston. And then when I reached Boston, the last flight to Manchester had already departed. I did not know anybody in Boston and had no idea where I would stay the night. A few minutes later, I saw a woman who was struggling with her luggage. She had been on my flight to Boston. I offered to help carry her luggage and put these heavy suitcases in her car. She thanked me and asked me where I was going. My English was not very good, but I managed to tell her my story and said I had to wait until the next day for the next flight.

She asked me if I knew anybody in the Boston area. I remembered meeting someone at the Kochi airport who told me he had an uncle who was a priest in Boston. He wrote the priest's address on a piece of paper and told me to visit him if I ever got the chance. When I told the woman about this priest, she told me that she lived very close to him and that she would be happy to take me there.

Though I did not know this priest, I decided to go to him that night. That woman gave me a ride to the priest's house and made sure I met with him. I thanked her for helping a total stranger like me, especially at night. The priest was very happy to see somebody from India. He welcomed me and gave me dinner and a room to sleep in. I told myself that I was going to be just fine in this great country, which had wonderful people in it like that woman and the priest.

After I received a master's degree in chemistry from the University of Massachusetts, I wanted to go on for a doctorate. A friend advised me that I would probably have a better chance to find a job if I broadened my knowledge base. I decided to switch my major to chemical engineering and received another master's degree in plastics engineering. I am glad I made that decision because it increased my exposure and knowledge in many areas of technology, and I believe it was a critical life decision that structured my future.

Keep Your Eyes and Ears Open for Opportunities to Invent

After that, I worked for a small company in Toronto, Canada, that made plastic cups and lids. Working in that industry was a total transformation for me. I soon realized that what I had learned in college textbooks might not be of much use in the world of industry. What was more important was *how* you learned. The training and experience you get during the learning process is exactly what you need to survive and succeed in any industry. The learning process you go through in college trains you to think, deal with problems, learn on the fly, analyze situations, and make the right decisions. That training will give you the confidence to do better in any field.

Textbooks are only tools for this type of training. I remember staring at my textbooks, which I kept in my office for years. I hardly picked them up or used them to look up information or for help with my work. A textbook may teach you how to solve a specific problem, but it's not a specific solution you are interested in; rather, it's how it taught you to solve that problem and how you can relate that to solving other problems. It is not what you learned; it is more important how you use your knowledge.

I got all the help I needed from the experience and training I received during my college days. I helped the company improve its machines for plastic processing, which led to a better yield. I also helped the company with its plastic resin selections. This company was strictly a manufacturing facility for commodity products without much focus on research. I soon realized that there was not much room for me to advance or to be very productive there.

The next year, I joined a larger company that made electrical insulation products. My first project was to develop a low-cost transformer that could last a long time when buried underground. The transformer was encapsulated in thick plastic that provided electrical insulation as well as impact, corrosion, and thermal protection when buried. This project was very exciting since it required design work on its chemical as well as its electrical aspects. My partner, an electrical engineer, handled the electrical design. To redesign the transformer, we had to work together every step of the way to determine the compatibility of the electrical and chemical components.

The formula used for the plastic mold they were using then was very costly. To redesign the chemical formula, I investigated each of the chemical ingredients in the existing formula in terms of its function, properties, cost, and ease of handling. The next step was to identify chemicals of the same family and compare them with those in the current formula.

After a series of tests, I came up with a new formula for the mold that called for fewer ingredients and process steps at a significantly lower cost and with improved performance overall. I was not familiar with some of these chemicals, but for the first time, what I had learned in college came in handy. My success gave me a lot of confidence. Working with an electrical engineer expanded my comfort zone beyond chemistry, and having the desire to explore areas beyond your comfort zone is an important element in becoming a successful inventor.

Every day, I walked the manufacturing area and stopped at each station. I talked with the operators mainly about the company's products but also about any issues and concerns they had. I also observed the overall operation and kept daily notes. The machines were old and the techniques they used were not so modern, but the production was smooth and nobody complained about the operation. Nonetheless, I saw a lot of opportunity for improvement by challenging the status quo. During the next couple of years, I proposed lots of changes in the manufacturing process and product that resulted in improved production and quality. The ability to challenge the status quo is a critical quality for any inventor.

I realized that the United States had much better opportunities for me to advance than did Canada, so I started to look for employment opportunities in the United States. While attending a seminar in Boston, I met with a research and development (R&D) director for a small company in New Jersey. He was impressed with the work I was doing for the Canadian company. He asked me if I would be interested in working for his company. I told him that I was but that I needed a permanent US visa, a process that could take years. He told me he would talk with the president of his company.

The next week, the president called me for a phone interview. After an hour of discussion, he offered me a job and said they would process my visa papers. I got a call from him every few months as our waiting game continued. A year later, I called the immigration office and learned that they had a backlog of such requests and that mine would not be considered for another year. I was very disappointed and conveyed the message to the president of that company. After not hearing from him for a few months, I concluded that the company had lost interest in me and that I could forget about moving to the United States.

But the next month, the president called and said that his company had hired a new CEO who had looked at my resume and wanted to talk with me. The CEO was visiting Toronto the next week and asked me to meet with him. We spent a couple of hours talking about my past and his new company challenges. At the end of the interview, he told me that he really wanted me to work for his company. He also told me to prepare to move to New Jersey in the next couple of weeks. I told him about my visa problem, and he said he would take care of that. To my surprise, I got my visa in two weeks, and my family and I moved to New Jersey the same month.

This company produced rubber molding for electrical parts and made electrical insulation tape using rubber formulations. As had been the case with my Canadian employer, this company had old and outdated processes and machines. My newly learned experience in electrical insulation really came in handy in my new place of employment. I closely observed every operation and all the products. Identifying issues and problems was my priority, and I again made many positive changes in their processes and products.

Identify the Opportunity—Take the Ball and Run with It

Identifying a problem can often lead to an invention. One of the problems I identified in this new company was that the current insulation tapes failed frequently at extremely high temperatures. These tapes were used to cover splices in high-voltage wires. If the tape was subjected to very high temperatures, it would melt or burn and expose the wires. That happened especially during lightning strikes. Based on this information, we determined that the tape needed to withstand an extremely high 3400° F for at least twenty seconds without compromising the insulation. Even ordinary steel melts and burns through when exposed to that level of heat. We all knew that this was an extremely difficult test to pass especially since any rubber or polymer formulations would burn and disintegrate at even much lower temperatures.

I was assigned to this project, and I had to work alone to solve this problem. For several months, I tested various formulations without coming close to the high-temperature requirement. But each failure was a learning experience, and I gathered valuable information as I analyzed each failure. A special flame-resistant rubber was used as the main ingredient, but it still failed as it burned and charred at elevated temperatures. I knew that rubber alone would not provide the protection I was looking for, so I tried a special filler material in the formulation. After mixing this special ingredient in the formula, I found that the resistance to burn-through and disintegration was significantly improved. Repeated tests showed gradual improvement but not to the level we were looking for. The research came to a dead end.

But then I changed the ratio of the ingredients, and the tests showed dramatic improvements. I was still a long way from our goal, but I was hopeful. I made some changes in the process conditions. To my surprise, I noticed that the process conditions for this formula were very sensitive and even slight changes in conditions could alter the properties dramatically. My new challenge was to identify the right process conditions that would create the optimum properties of the formulation. Many trials later, I identified the optimum conditions and realized I might have a chance to succeed if I could manipulate the right chemical ratio and the right process conditions.

I made more samples for the final test using plasma arc test equipment that could handle temperatures equivalent to that of a direct lightning strike. At that time, only the University of Connecticut had such equipment in our area. The test was very expensive due to the high equipment cost. When I called the professor who was in charge of this test equipment, he asked me what type of material I was planning to test. When I told him that it was rubber tape, he discouraged me from moving forward with the test and told me I would be wasting my time and money. The comments from the professor made me very nervous, and I began to lose confidence. But I had worked hard on this project and was not going to give up. I eventually convinced the professor to do the test.

I reached the university with my test samples on a Monday morning. The rubber tape material that I had made was about three inches wide and about a quarter-inch thick. He said that most of the material he had tested including some metals had failed miserably. He said that only a handful of materials he had tested had survived twenty seconds of exposure to the plasma arc heat source.

The professor wrapped the tape on a lead pipe and secured it on the test jig. Lead melts at a low temperature, so if it melted, the tape that was supposed to have insulated it would be considered a failure. The sample was subjected to the plasma arc heat source; the tape started to burn immediately, but that stopped in a few seconds.

As the test continued, we noticed a strange crust forming on the surface of the rubber as the result of the reaction between the rubber and my special ingredient under the extreme heat. The professor was curious about what was going on.

The test went on for more than a minute, and still there were no signs of failure. I felt good for having passed the twenty-second requirement. The professor decided to continue the test for another five minutes. As there was still no sign of failure, he started to suspect the test equipment. He looked the equipment one more time and made sure everything was running right. He told me that he has never seen anything like that; he wanted to continue the test.

After twenty minutes of exposure to the plasma arc, my test material still held on without failure. The strange part was that the machine started to overheat since it had never run for such a long period. The professor had to turn the machine off due to that overheating.

After examining the piece of rubber tape that survived the test, I determined that the initial burning reaction had created a hard crust similar to ceramic that was resistant to very high temperatures. The crust could resist the plasma arc flame and prevented it from further degradation and penetration through the material. To our surprise, we also noticed that the lead pipe surface below the quarter-inch thick tape had not melted, which meant the newly formed crust had excellent thermal insulation properties. This unexpected result was even more significant, and it broadened the scope of my invention.

I could have easily given up on this project due to my lack of progress, but my ambition to succeed and my persistence had led me to try different and special materials and unusual process conditions.

High-Voltage Insulation Tape Exposed to 3400° F

I was excited; I called my boss to give him the good news. When I got back to the office that afternoon, everybody in our department and the CEO were waiting to congratulate me. The CEO took me to the side and told me, "There was a good reason for me to hire you." During the next week, he arranged a cash award for my successful project completion.

We had customers eagerly waiting for such a product, and sales started to take off. Initially, the company wanted to apply for a patent, but the decision was that we would keep this a trade secret since the company was convinced that due to the complexity of the formula and the process conditions, it would be very difficult for the competitors to copy. I generally agreed with that decision, but for me, it was a big loss; it would have been my first chance to get a patent.

We started to look for other applications for this product with unique temperature-resistant and thermal-insulation properties. During that time, I read about the space shuttle having problems with the tiles that covered the outer surface. Some of the tiles were burned through exposing the body to higher temperatures during reentry. We contacted NASA to discuss this new product as a possible solution for the required heat resistance during reentry. We had some initial contacts with NASA, but unfortunately during that time, the second shuttle blew up and NASA terminated the shuttle program.

That was my first official invention, and to this day, I consider it my most significant invention. I keep the tape that survived the plasma arc test in my office and look at it every day with pride.

Moving On

My wife, Mary, our daughters, Charlene and Crystal, and I bought a house in New Jersey. Mary, a psychiatrist, started a private practice. The following year, I found out that our company was being bought by a larger company. A meeting was called to make the announcement about the new company acquisition. Officers from the new company announced that they decided to close down the operation in this area and move to it Massachusetts. One of the top officials told me that the company was offering positions to only 7 of the 150 employees at the New Jersey company. He told me I would be the first to be offered a job at the new place. I was flattered but definitely not happy.

After discussing this with my wife, I told the president of the new company that I wasn't interested in moving. He made me a better offer, but I said that my wife's situation was the main reason for my decision. A week later, he told me that if I took the job, he would find a position for Mary at a hospital on whose board he served. I was very surprised to hear something like that, and I thanked him for his generous offer. After a long discussion with my family, we nonetheless decided not to make the move; I would look for local opportunities.

I sent my resume to a recruiter, and the next week, I got a call from a company called Sealed Air Corporation not far from us. I had never heard of that company before and had no idea what it did. In those days, we did not have the Internet, and I did not have much time to look up this company. I initially thought it had something to do with air conditioning, but it turned out that the company had invented Bubble Wrap. This simple product fascinated me, and I always wondered how it was made.

I had a long and interesting interview with two vice presidents; they were impressed with my resume and asked many questions about my work experience. I also learned about their expectations for me. They told me the vice president of manufacturing would interview me before they made a decision.

In two days, one of the vice presidents who had interviewed me told me that the third VP would be out of town for some time. He said they had briefed him on my resume and the discussion we had during the interview. I was thrilled to find out that they had already made up their minds and had decided to make me a job offer.

I was lucky that the third vice president had not interviewed me since he probably would not have hired me. I found out later that he was a fan of a new polymer resin that had been introduced to the market very recently and his top priority was to use that resin in the Bubble Wrap formula. I had never heard of that particular resin before, and I would have failed badly in that interview.

The Sealed Air Corporation was a much bigger company than was my previous employer. For the first time, I got an office of my own with a window and a secretary to help me with my work. I was impressed.

On my first day, someone introduced himself to me using his first and last names and asked me how I liked working for the new company. He told me that he had heard a lot about me, and he talked to me as if he knew me well. He also asked about my family as if he knew all about them as well. After spending about thirty minutes with me, he told me he was pleased I had decided to join the company.

After he left, my secretary came rushing into my office and asked me if I had recognized him. I told her that I just learned his name but didn't know what he did. She laughed and told me he was the CEO of the company; she said he had asked her for all the details about me before coming to my office.

Our facility had a fully equipped laboratory, and I had a lab technician to help me. For the first few days, I reviewed all the projects assigned to me. I visited the manufacturing area to learn all the processes and machines for making Bubble Wrap. My boss told me that I didn't have to write a monthly report for that month since I had been there for only a week of that month, but I submitted a four-page report based on my observations during that week anyway. After seeing

THE INVENTOR IN YOU

the long report, my boss asked me how anybody could have written a four-page report after having worked for just a week. I politely told him, "Don't forget that I came from the rubber industry—I learned how to stretch things."

The more I learned about the company and the people there, the more I liked it. I felt right at home there. I decided to make this my second home and a place to work until I retired. After five years there, I was promoted to the position of director of R&D. I retired this year after thirty-four years there.

It's Not What You Have Learned but How You Use Your Knowledge

My first project was to introduce a new polymer resin into the bubble formula. As I indicated earlier, this was a top priority for the vice president of manufacturing who luckily had not interviewed me. I started to learn about this new resin, which was very new to the industry. At that time, not too many processors had much experience in handling this resin. Our company wanted to get a head start with this resin before the competition did. It had many benefits including improving the properties of the product and lowering the cost.

Since we did not have much experience in handling that particular resin, we conducted thorough research on it and conducted a series of trials. The initial trials were disappointing, but we learned a great deal from those failed trials including the fact that for this new resin to process properly, we needed to blend it with some other resins. After six months of research and many trials, we finalized the formula and process conditions.

The new Bubble Wrap formula using the new resin was commercialized during the following month. I used a systematic approach; I carefully planned each step of the evaluation, learned from failure, and was not afraid to take chances. All these are the key elements of a successful invention. Unfortunately, we could not get any patent protection for this formula or process as there were a few related patents. As you will learn later on from this book, not all inventions can be protected by patents. But the next week, I got a call from our CEO congratulating me for commercializing Bubble Wrap using this new polymer formula.

Avoid Tunnel Vision and Think Outside the Box

While working on the above project, I was also given another challenging project. The company had just introduced a flame-retardant Bubble Wrap. Within weeks, we started to get complaints from customers. They basically rejected the product stating that a dusty residue was blooming out of the Bubble Wrap and contaminating their products. We had to identify the cause of this problem before we could try to resolve it.

After lab tests, we determined that the one of the key ingredients was not compatible with the polymer formula and it leached out of the Bubble Wrap film over time, especially at high temperatures. We knew what caused the failure, so the next challenge was to find a solution. The work continued to reformulate the flame-retardant recipe. We studied the chemical nature and properties of this key ingredient and tried to alter it to make it compatible with the polymer resins.

We tested many combinations using multiple ingredients in varying ratios without any positive results. I then realized we were focusing only on the obvious materials for this exercise. Thinking outside the box, I started to test a few chemicals that we had normally considered ineffective in such formulations. I decided to combine one of the materials from this new list of chemicals with the key ingredient, and the result was a total surprise. The film made from this formula did not show any signs of blooming, and no dusty residue was visible even at elevated temperatures. This new formula was tested in the manufacturing process. The Bubble Wrap product underwent extensive aging and elevated temperature tests without showing any signs of the dusty residue. Our company was back in business selling flame-retardant Bubble Wrap.

I was very excited about this unexpected result. Our R&D department recommended that I file a patent application for this new Bubble Wrap formula and process. The patent application was filed, and I got my first patent for a flame-retardant Bubble Wrap.

Always keep an open mind during the research and development process; avoid tunnel vision, and think outside the box. These are very important tools you can use when you hit a roadblock in your development process.

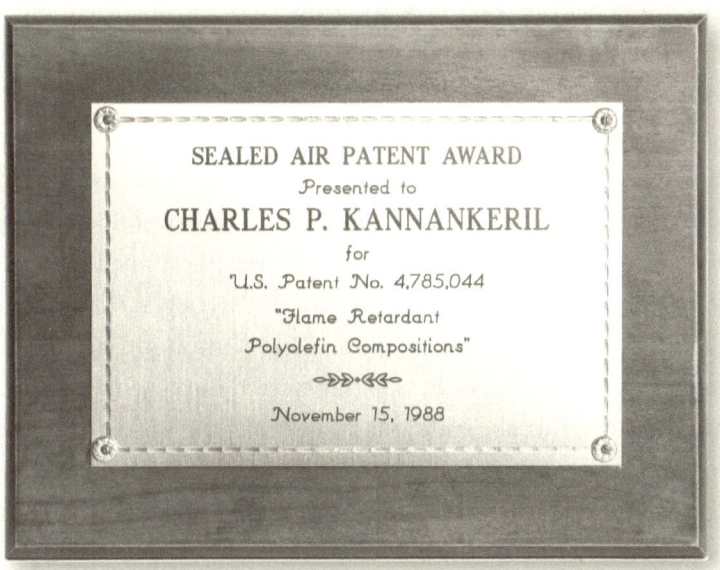

Flame-Retardant Bubble Wrap Patent

Look at the Big Picture and Find a Creative Way to Invent

Our company used to make those absorbent pads that go in meat trays to soak up what's called meat purge—blood—during shipping, storage, and store display. The goal was to improve the absorption capability for the pad and retain the fluid without its getting squeezed out when a heavy piece of meat was placed on it. We tested many options with varying results, but none of them was a fully acceptable solution. I realized there were limitations to how much additional absorption we could get without the fluid being squeezed out. The project came to a standstill.

We decided to revisit the problem and possible solutions. The customers wanted increased absorption and retention of fluids, so we were focusing on the pads. When we looked at the bigger picture, we realized that the pad was only one way to improve absorption; the real need was to improve the absorption of the tray. By thinking outside the box, we tried to redefine the problem and the goal. These absorbent pads were made separately and placed in the tray during packaging. For the meat packer, that was additional cost in material and labor. My thought was that instead of trying to improve the pad, why not eliminate the pad and make the tray absorb the fluids? I kept that thought in the back of my mind to work on.

Several days later, my older daughter, Charlene, was celebrating her fifth birthday. We invited all her friends to our house for a party, and I hired a magician to entertain the guests. I paid close attention to one magic trick he performed. The magician poured a glass of milk into an open bottle. The bottle was painted black, so nobody could see what was inside. When all the milk was transferred into the bottle, he asked my daughter to pour the milk out by turning the bottle upside down. When she tried, nothing came out of the bottle, and it looked like the bottle was empty. I was intrigued by this magic trick and tried to figure out how he had done it. Since a magician will never reveal his secrets, I had to find the secret myself.

After many hours of frustration, I figured out that there was a funnel inside the bottle and sealed to the mouth. You can pour anything into the bottle through the funnel. Since the funnel tip extended all the way to the bottom of the bottle, the milk got trapped between the bottle and the funnel without spilling when the bottle was turned upside down.

I immediately made a connection between this magic trick and the problem I was working on. My logic was that if I could somehow make the fluid go into a meat tray and get trapped inside without coming out even after turning it upside down, I could create an absorbent tray without using any absorbent pads.

I was confident that I could come up with a new tray design from what I had learned from this magic trick. If you stretch your imagination, you can visualize a meat tray with a hole in the center that can act as a funnel. Any fluid poured into this tray will flow to the center hole and out. I placed another tray at the bottom of the funnel tray and sealed both trays together along the lips. I poured some water into the top tray. All the water disappeared through the hole. Since the tray was opaque, nobody could see where the water had gone; it was trapped between the two trays and did not spill out when the tray was turned upside down or even after shaking it. I knew I had something great going. If successful, this idea would be a better way to retain the fluids in a tray, and most important, it would eliminate the use of absorbent pad altogether.

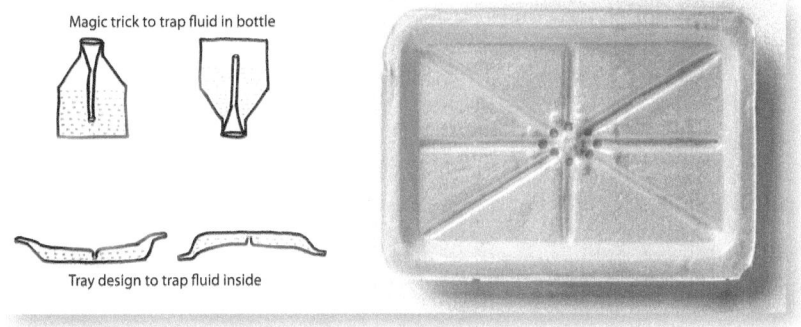

Meat Tray Design—Magic to Reality

We made several modifications to improve the flow to the center, including creating flow channels on the top tray and increasing its slope. After I was confident with the first working prototype, I wrote the details of this invention in my monthly report. A few days later, I saw a copy of my report back on my desk with comments in green ink. I knew immediately whose writing it was since only our company CEO wrote with green ink. He had written "Excellent!—Please call me when you get a chance regarding this new invention." I called him right away, and he said he wanted me to present this invention to the next board meeting.

I was a bit nervous since I had never been to a board meeting. I prepared all the samples for my show-and-tell and made presentation slides for the overhead projector (no laptops back then). The presentation went very well. All the members were impressed and congratulated me.

One board member happened to be the inventor of the absorbent pad. After my presentation, another member joked that this invention would make the first member's pad patent obsolete. The member who was the inventor joked that he would be willing to pay me to go back to my country and forget about this invention. He shook my hand, congratulated me again, patted my shoulder, and told me to keep up the good work. The CEO said I could count on the board's full support for this project.

That meant the pressure was on me. I kept making modifications so we could make all the possible claims we could about this new product while preparing for a patent. Two years later, we received the good news that the patent office has decided to allow all the claims and a patent would be issued shortly. A month later, I was shocked to receive another letter from the patent office stating that a final international search had revealed that a similar patent application had recently been granted in Netherlands and I had missed the filing date by a mere three months. Earlier prior-art search done by our lawyers and patent office had not revealed the work done by inventor in Netherlands. It was a total shock and disappointment for all of us. I had never expected something like that would happen. It is interesting to note how people think alike in different parts of the world and come up with similar idea without any prior knowledge of each other's work or having any contact with each other at all.

Often, an invention is the result of something triggering your creative mind. You probably have several unsolved problems stacked up in your brain. Every time you see something different or strange, your brain can shuffle through the stack and try to relate to this new event. Eventually, it will hit a match that can lead to the solution to your problem.

Improving Products Using Other Technologies

We continued to look for ways to improve the pad's absorption capabilities. At that time, my daughter Crystal was still a baby who used diapers. One day, it was my turn to change the diaper she had slept in, which was soaked, but I noticed that it did not leak. I cut it open and noticed a very thick gel in the diaper. I was curious about that and decided to do some research. It turned out that this material was known as a super-absorbent polymer (SAP) used in diapers to increase their absorption capacity. This powdery material can absorb up to a thousand times its weight in water and can retain it as a gel. I made an immediate connection to the absorbent-pad project.

One would think that using this SAP powder inside the absorbent meat pad would be an easy solution for improving and retaining the absorption, but things weren't that easy. We found out that because of chemical differences between meat purge and urine, the absorption capacity of SAP was much lower for meat purge than for urine. But after some more research, we optimized the absorption by selecting the right SAP particle size. The pad design and the machines were modified to incorporate the SAP.

A new challenge was that though the SAP was used in diapers, we needed approval from the Food and Drug Administration (FDA) because these new pads would be coming in contact with food. Getting that FDA approval took us more than six years!

We finally commercialized the new SAP absorbent pad for meat-packaging applications, but we could not get a patent for this invention. The patent office cited similarities between our pad and diapers even though it would be used for a different product and purpose. I learned that patent officers are very stringent, and it is very difficult to get around the prior art and get a patent. (I will discuss the patent-application process for protecting an invention in a later chapter.)

Regardless, I continued to work on the absorbent pads and made several modifications to improve their ability to absorb and make it more competitive, and we were ultimately granted three patents for the modified pads.

Invention by Improving a Product for Other Uses

While trying to improve the absorbent pad, we tried to explore ways to increase its sales. The market for this product sale was limited, so we looked for other ways the pads could be utilized. We realized that they could soak up the grease bacon gives off when it's fried. The challenges were that the current absorbent pad could not withstand the cooking temperature and that it required additional testing to comply with FDA regulations.

We identified a high-temperature polymer film, but the adhesive was still an issue. We decided to eliminate the adhesive and focus on joining the film by heat-sealing it. Unfortunately, the type of film we tested did not heat seal to itself. Continued research showed that a similar film had been developed by the polymer industry; one side had been altered to make it heat sealable. That was exactly what we were looking for, and we completed the product development in no time. The use of this special film and the sealing method to produce an absorbent pad for high-temperature cooking applications were unique, and we commercialized it successfully. We were granted a US patent for this product and process.

In spite of the roadblocks they face, inventors who are persistent will come up with successful inventions.

Invention to Meet Customer Needs

Trying to fulfill a particular need often leads to an invention. During the initial stages of the AIDS crisis, there was a general concern about handling body fluids, especially among the medical personnel who tested specimens. Many times, specimens leaked out of the package during shipping and contaminated the surroundings. Our company was contracted to produce leak-proof packaging for shipping these specimens, and I was given the responsibility of developing solutions.

Our company was already producing absorbent pads used in meat trays, so I decided to take advantage of that technology. I redesigned the pad structure and created a leak-proof outer bag for safely shipping biohazard fluids. The new design consisted of an inner fluid-permeable film, a thick and rugged outer film, and absorbent materials in between them. Besides the original request, the final design had many value-added features. Disinfectant chemicals were incorporated into the absorbent medium to neutralize the fluid in the event of leakage from the container. Other features included insulation to protect the specimen from extreme temperatures and cushioning to prevent product damage during shipping. These additional features were well received by our customers.

We had the vision, and we took a proactive approach to solving customers' problems. Our customers were very pleased with all the additional features, and the company enjoyed a very good margin for this product. The same year, I received a US patent for this invention.

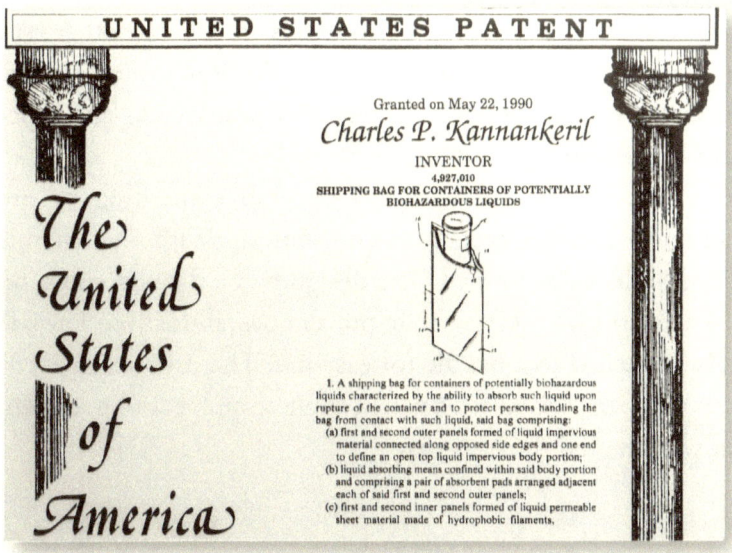

Patent for Biohazardous Fluid Shipping Bag

This was a good example of successful market-driven product development. The marketing team has already done a market analysis and requested the R&D for the product development. The R&D anticipated the issues and took a proactive approach by providing a better solution. Before developing the product, we in R&D put on our customer hats and identified potential issues by thinking like a customer.

One way to increase your customer base is to provide value-added products. Some companies design and make what they were asked to make—no more and no less. But if you think like a customer, you can come up with a new wish list for the product—user friendly, long lasting, additional features to meet other needs, bigger, smaller, recyclable, reusable, degradable, and so on. Each time you think like that, you will find new opportunities to invent something.

Research-Driven Invention

During the same time, I started to think about the infectious-disease crisis and to look for potential areas to address the contamination issue. I thought about people using cotton balls to wipe blood from body surfaces. Medical professionals always use gloves to protect themselves, but such gloves are not always available in other areas and especially in most households.

The opportunity here was to develop something to prevent contamination under certain circumstances. With that in mind, I developed a blood-wipe pad. The design had a flat absorbent surface attached to a thick backing film to prevent contamination. The backing film also attached to a handle for easy use. The product worked well, and our company submitted a patent application for this invention. The patent was granted the next year.

Patent for Blood-Wipe Pad

I visited a potential customer for the blood-wipe pad with our marketing personal, and the customer was impressed with the product. Unfortunately, this customer dealt with all types of medical supply products and wanted to buy products only from suppliers specializing in these types of products. At that time, our company had never produced any medical-related products, and we had had no experience in marketing them. In spite of receiving a patent, this product was never commercialized for many reasons.

It was a typical example of a research-driven product. I started to develop this product without many interactions with the sales and marketing team. At that time, I did not have any idea whether there was any real market for such a product or about the distribution channels.

You may think that you have created something great, but you aren't going anywhere if you cannot sell it. Inventing a product and marketing it are different tasks. These two aspects need to be tied together closely for any invention to succeed.

It's important to conduct a good market analysis so you can identify the market size and volume before you embark on any product development. It may be okay for independent inventors to skip this stage since they can approach the right company specializing in similar products.

Market-Driven Invention

Often, customers come to you with a problem they want solved. The sales and marketing staff do their homework based on the market potential, company core competency, financial impact, and most important, the R&D resources. Based on the outcome, they decide to accept or reject a proposal for a new product. Acceptance makes for a great opportunity for an inventor to jump in.

During the crisis involving anthrax in the US Post Office, our company was contracted to come up with a solution to keep the postal system free from such chemical contamination. We formed a team to address this issue. I took a leading role in this exercise and came up with a few design proposals. One proposal was for a specially designed bag that could be placed inside each post office mailbox to serve as a liner that would keep postal employees from being contaminated.

I wanted to find out the dimensions of a typical mailbox to finalize the design. Instead of waiting until I went to the office the next day, I decided to go down the street to a mailbox and take some measurements. While I was taking my measurements, I saw a police car coming my way. I was a bit shaken up since the police were still looking for whoever had mailed the anthrax-laden letters. I was afraid I would look suspicious, so I put away my tape measure and pen and paper and walked the other way. The police car drove by; apparently, the officer did not even notice what I had been doing, but that was definitely a scary moment for me!

In a very short time, we came up with several designs to address this issue. The team looked at various designs and selected the top two. I was very proud of the fact that both the top designs were my proposals. The company decided to apply for patents for both of them, and two years later, we received them.

We never commercialized these products since the crisis was resolved and the post office no longer needed them. Regardless, my team was very proud of having addressed an important crisis facing our nation, and we had the satisfaction of coming up with a solution quickly.

An inventor should always be ready, willing, and able to face challenges and find solutions in a timely manner. Having the confidence to solve problems and deliver the results before the deadline is the key for an inventor to survive in any industry.

Invention by Combining Technologies

When I started working for Sealed Air Corporation in 1983, it was not a very big company, but company sales hit the $100 million mark that year. The two main products the company made were Bubble Wrap and InstaPak foam. Bubble Wrap is made by thermoforming—applying heat to a film to get the bubble shape on one side and sealing it with a flat film on the other side; the air trapped in the bubbles provides cushion. InstaPak is a foam that can be created on demand at the customer site. Our company sells two different liquid chemicals along with mixing and dispensing equipment to the customers. The customer mixes the chemicals and dispenses the foam to create a protective cushion around an object prior to shipping.

I was introduced to these products during the first week at the company. After studying these products, I started to wonder if there were ways to improve these products or create something different using these technologies. Popping Bubble Wrap is fun, but not knowing much about its properties and how it performed in the industry, I felt that it could be vulnerable to pop and fail when heavy objects were packaged in it. One good thing about it is that it is highly flexible and customers can easily wrap it around objects. On the other hand, InstaPak foam could provide a durable cushion protection for heavy objects though it is not very flexible and not as easy to use. I thought it would be nice to have these positive features—high flexibility and durable cushion—in one product. My first thought was to fill the bubble cavities with the foam; I thought that would create a heavy-duty cushion product that provided flexibility as well as a durable cushion.

At that time, our company had mainly two divisions, the Bubble Wrap division and the InstaPak foam division. Both divisions operated pretty much independently; they did not regularly interact. I was not sure how to proceed with this idea without interfering with the foam division. But the next day, I presented this idea to the R&D vice president. I felt good about this idea and was expecting a positive reaction from him. To my surprise, he laughed and asked, "Don't you know that air is free and foam costs money? Why are you even thinking of replacing air with foam?"

I was shocked to hear that from him, but it made me think. He had brought up a good point. I was not sure there was a market for such a product or if customers would pay more for it. I should have done some market analysis to find out if there was a need for such a product or if customers would pay more for it. If I had had the answers to those questions, I would have reacted differently. I learned an important lesson. Apparently, the market was not ready for such a product at that time.

It is strange that ten years later, our company introduced a product very similar to my original suggestion. Over the years, the market grew; people were shipping more and different types of products. Of course, there was a new need for a heavy-duty cushion and a material that was flexible enough to wrap around objects. Without knowing the proposal that I had made years ago, somebody in the foam division came up with a similar product idea to meet the current market need. I was thrilled to see the commercialization of such a product years later. Having had that vision many years prior to that made me very proud.

It is very common to come up with an idea for a product or a modified version of an existing product without knowing the market need; there's nothing wrong with that—it's still an invention. You cannot judge an invention's worth by current market need. Who knows? Maybe you had a great vision of the future and you were ahead of current market trends. It all depends on what area you're working in.

Keep in mind that such inventions won't go well if you're working for an industry that is always trying to keep up with market demand; some companies don't have the time and resources to invest heavily in the future. Besides, they may not want to take chances and test your idea, especially if their current products are doing well and they don't want to make any changes that might backfire. You just have to wait until the right time for the market to accept your idea.

Independent inventors don't have to worry about such concerns; they can invent things for the future and try to sell them to a company. A company could be interested because it could get an invention without development costs and could slowly market it or simply keep it safe from the competition.

Having a Vision and Inventing for the Future

Bubble Wrap is handy, but it is bulky. That makes it expensive to ship and then store in warehouses. Fortunately, during the early days of my employment, our company was enjoying a profitable business with no significant competition. At that time, our customers were not concerned about storage or shipping costs, so my company was comfortable with the situation and was not ready to even consider any changes. But I believed there was a better way to make Bubble Wrap and supply to the customers.

I proposed a bubble structure that could be made without air inside and shipped flat to the customers, who would inflate it on demand. To test the concept, I made prototype bubble samples. The first was for large square bubbles made by sealing two pieces of film together to get the desired bubble shape. The product had parallel rows of bubbles, and the bubbles were connected to each other in each row. Each row was connected to a main channel, which was capable of supplying air to all the rows during inflation.

One would think that that was a good start and that the company would embrace the idea and support this project. But again, my company was the only major player in the Bubble Wrap field without any serious competition; it was enjoying a good business, high profits, and no complaints from customers. Gas prices were low, and there were no concerns about shipping and storage costs. The company decided not to entertain the idea at that time, and nobody argued with that decision. For the next fifteen years, I occasionally brought this proposal to management but always got the same answer.

But things changed over the years. The competition became stronger, prices became very competitive, and gas prices skyrocketed. Management saw the opportunity for change and decided to work on the inflatable bubble products. We created a team to develop the product I had conceptualized.

Besides our R&D staff, we had good representation from the sales and marketing, manufacturing, accounting, and legal departments. During the development process, we made sure we covered many features to make the product very competitive. We made many new claims for the patent applications. The product was launched successfully, and we were granted three patents for this new product.

In a business world, it is always good to study the market before you try to invent a product. If the market is not ripe for a your invention, it may have to sit on the shelf for years before it can be commercialized successfully. But sole inventors may have a different priority—to be the first one to invent something rather than waiting and losing the rights to it. On the other hand, if you invent and patent your invention early but don't use it, you may lose valuable time; your exclusive rights to your patent will last for only so many years. It's double-edged sword. I will provide more information later in this book on how to protect your invention.

Don't Discount Anything—The Answer May Be Right in Front of You

In R&D, we always think up different ways to use existing products to create a value-added product. Bubble Wrap is usually taped up after it has been wrapped around something. I visited one of our customers and saw people struggling to secure the Bubble Wrap by using shrink-wrap, a plastic wrap that is heated to shrink a bit and secure whatever it's wrapped around. I wondered if there was a better way of securing Bubble Wrap. I couldn't come up with any solutions immediately, so I put the thought in the back of my mind.

The next week, I was testing some shrink-wrap. After heating it, I got the film tightly wrapped around the product. That triggered an idea. I visualized a sheet of Bubble Wrap between the product and the shrink-wrap to provide a nicely secured bubble cushion around the object. My first thought was to attach shrink-wrap on one side of the Bubble Wrap.

Then I had to figure out a way to attach the shrink-wrap to the Bubble Wrap. The choices were to use an adhesive spray between the bubble and film or to heat-seal the bubble and film together. Neither was a good option; the high cost of adhesive spray would have made the product too costly, and the shrink-wrap would shrink prematurely during the heat-sealing process.

We nonetheless decided to revisit the second option of heat-sealing the film, but that would cause the shrink-wrap to tighten up then, and it would not regain its ability to shrink again when that was required to seal and secure something. I looked at the shrink characteristics of this film and determined the temperature at which the film would shrink. I then looked at the softening point of the Bubble Wrap. After close analysis, I noticed that the film could be softened and sealed to the Bubble Wrap at a temperature a few degrees below its shrinking temperature, so I took advantage of that narrow temperature range.

We tried to modify the equipment to maintain this narrow temperature range and by doing so; we produced the first shrink Bubble Wrap by heat-sealing shrink-wrap to the bubble sheet without any premature shrinkage. This gave our customers a value-added product they could apply easily and secure the product better. I was granted US patent for this shrink Bubble Wrap product.

Don't get discouraged when you hit a roadblock during a development process. Reevaluate the situation, dig deeper, and find out exactly why something doesn't work. The answer may be right in front of you. Don't discount anything; there may be a nonobvious solution.

Helping Other Divisions through Brainstorming

I was frequently invited to brainstorming sessions with other divisions to get a different take on their projects. One time, I was invited by our InstaPak division, which makes cushion foam that activates on demand when two chemicals are mixed. The new project was to develop a bag with a small amount of these chemicals that could be mixed by hand and produce foam on demand. The challenge was to seal these chemicals separately inside a small bag without mixing prematurely and break the seal on demand to mix them together.

The main purpose of this session was to come up with a seal strong enough to keep the chemicals apart and at the same time weak enough to break with just hand pressure. During this session, one of the ideas that I suggested was to create a seal that was easily breakable by printing the adhesive in a special pattern. At the end of that brainstorming session, that was selected as the top idea. We received a patent for that, and it is the currently used in this product. And it became one of the most successful new products for that division.

Getting Ahead of the Competition by Identifying Their Weakness

The bubble division was well equipped with plastic processing machines; we made several types of films for bubble as well as nonbubble products. The company wanted to expand the business to other product areas. One of the suggestions was to make small bags to pack produce in supermarkets. There were many manufacturers who supplied these produce bags in rolls, so we wanted to supply these customers with something better. At that time, another supplier came up with a compact bag roll by folding the bags prior to winding. This supplier also had a dispenser to release one bag at a time. We looked at this design and found many issues with the roll as well as the dispenser. I was asked to take the lead to come up with a better design and to address the issues with the current bag.

Identifying the issues correctly is the key to solving problems. If you don't understand the nature of the problem, you won't be able to fix it. We spent several days in supermarkets observing shoppers to determine how they used the product. We listened to their complaints about the product and asked them what they would like to see improved.

They were used to pulling the full-width bag from a roll using both hands—one to hold the roll and the other to tear off the bag. The new design had a compact roll less than half the width, and customers could separate a bag by pulling with one hand. The bags were narrow and compact on the roll but easily unfolded into a full-size bag after separating from the roll. It was interesting to observe the reaction of the customers when they tried to use these for the first time. Some first-time shoppers didn't unfold the bag; they just threw it away thinking it was a narrow-sized bag.

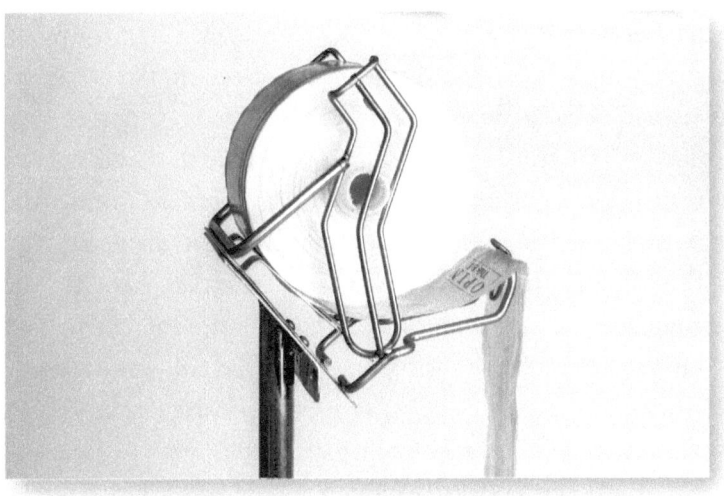

Produce Bag Roll and Dispenser

We used customer feedback when we were designing that product, and we came up with three designs for which we received patents. We successfully commercialized the product by the end of that year.

You can come up with many features in a product, but that won't do anything for you if your customers have no use for them. By the same token, you need to focus on the actual customer need. The key is to address the features your customers are interested in. Early customer feedback is critical for developing the right products.

Solving Customer Problems

Our supermarket customer pointed out an issue with the new design for the bag roll and the dispenser—some shoppers were taking the new compact rolls of bags off the dispenser and taking them home! The store owner wanted a redesign to stop that from happening.

The bags were wound on a plastic core that stuck out about an inch on both sides of the roll. The dispenser had a channel to accommodate these extended core ends. Once the roll was placed on the channel, it rode up and down freely. I thought about ways to stop the upward motion once the roll was seated in the channel.

One day, I was playing with a lock and key I had in my office. I noticed that the key entered the keyhole and would not come out after it was turned. That gave me the idea that solved our current problem. I decided to test the key and keyhole combination on this project. After testing a few prototypes, I created two notches on the core ends and a pair of matching tabs at the entrance of the dispenser channels. Once the roll was placed on the top of the channel, it rotated until the notch matched the tab. The roll then could enter the channel and was free to move around without coming out. The roll was secured inside the channel, and the product was ready to use. The notch and tab combination made it very difficult for any shoppers to steal the rolls. This design was successfully commercialized, and we patented it.

Many times, you will struggle to find an answer to your problem. Always keep that problem in the back of your mind and try to relate it to anything you see or do. By this simple practice, you will be surprised at how many clues you will find that will solve your problems.

Expanding the Business Using Core Competency

One of the many ways to expand a business is to look at its core competency—what it does the best—and look for related products that can be manufactured and marketed relatively easily. One such product we identified was a drainage board used as a protective wrap on underground walls to prevent basement flooding. The idea was to collect and channel away water from around the walls and prevent it from entering the building. The product had a rigid flat panel with bubble-shaped protrusions on one side. The bubble tops were attached to a geotextile fabric. The flat side went against the wall and the fabric side against the dirt. Any water in the dirt was filtered through the fabric, channeled to the bottom, and guided away from the walls through large pipes. This product was already on the market; but we thought we could compete in this market with a better product.

Knowing the similarities between the panel and the Bubble Wrap structure, we thought we could easily make the panel using our existing plastic-processing equipment. Modifications were required to produce rigid panels compared to thin film used in Bubble Wrap, but after a few trials, we came up with a product similar to the existing panels.

After studying the product and its use, we found out that the existing products were limited as to how deep they could be used underground. The deeper a foundation went, the greater the pressure on it, and that pressure could collapse the panel structure. We proposed two different approaches to increase the structural strength—increase the product thickness or use a different type of polymer. Our tests showed that we had to increase the thickness so much to obtain the required structural strength that the cost became prohibitive. And we found out that the new polymer was weakened by chemicals in the ground. We had a very interesting challenge—balancing the strength, chemical resistance, and cost of the panel. That was definitely an opportunity for creative minds to come up with a well-balanced product.

After several trials, we decided to use a combination of polymers. We made a laminated panel with the core as the new polymer for structural strength while the old polymer was used as the outer skin for chemical protection. By balancing the amounts of polymers used in each layer, we were able to keep its costs acceptable. The design was finalized, and the product passed the structural-strength and chemical-resistance requirements. The product was successfully patented and commercialized.

Drainage Board using Bubble Design

When you're developing a product, you will commonly come up with many ways to solve partial problems. What I mean is that rarely will you find one solution for all your problems. But instead of discounting these options, you may find the perfect solution by combining some of those partial solutions. In the above case, the old polymer was good for chemical resistance but weak on structural strength, and the new polymer was structurally strong but weak against chemicals. Instead of discounting both these polymers, we used them in tandem and made a perfect product by taking advantage of each polymer's good qualities while overcoming their bad qualities.

Invention from Acquired Technology

Over the years, our company grew exponentially mainly through acquisitions. Each time we acquired a new company, a good number of new technologies came with it, and that often led to new inventions because we could combine and modify old and new technologies.

We became familiar with products and technologies belonging to the new company, and that gave us a chance to take a fresh look at those products and technologies. When outsiders look at an existing product or process, they may see things differently and most of the time with a positive twist. Maybe the people who operated the previous company were enjoying very successful products and saw no reason to improve or change. Getting a new team to look at things from a different angle is very much like thinking outside the box.

One of the acquisitions we made was a company that made bags that would show evidence of any tampering with them. One of the main products was a durable bag used for carrying sometimes a lot of cash for night deposits to a bank. The problem was that dishonest people could steal money from these bags, so this company's tamper-evidence bags deterred that. But of course, crooks were working equally hard to get around the tamper-evidence features as soon as the bags were introduced. That had become an ongoing struggle when we acquired this product line.

Some of the early technologies the new company had developed included the use of a strong adhesive to close the bag so no one could open the bag without destroying or tearing the outer film, which would have indicated the bag had been tampered with. But thieves found a way to open the bag without destroying the film and close it without any evidence of tampering; they gently heated the adhesive strip on the closure flap with a hair dryer until the adhesive was softened. The flap opened easily when the adhesive was warmed that way. The money was removed, and the flap was closed. The adhesive became as strong as before after it had cooled back down— no evidence of tampering.

The previous company had spent several months trying to meet this new challenge. One option was using adhesive that didn't soften under anything but extreme heat, but they couldn't find anyone who made that. They decided to "stick" with the same adhesive but with a new wrinkle—they printed heat-sensitive transparent ink on the flap that changed color and left a clear "VOID" message when heated.

But of course, some thieves got even more creative. They used a cooling spray can with compressed air to freeze the adhesive. The extremely low temperature tends to make adhesive less tacky, so they could open the bag and steal money. As the frozen adhesive warmed up to room temperature, the flap was closed without any evidence of tampering. It was a cat-and-mouse game.

One way to overcome the new threat was to develop a specially designed print between the adhesive and the film that became visible only when the flap was lifted, and a print technology was developed to do that. The printing was done in multiple steps, and it left an opaque, white film surface. A clear "VOID" message would appear on the flap if any attempt—hot or cold—was made to separate the flap.

But the game continued. We had another report of tampering that involved licking the adhesive prior to closing the bag. The adhesive became weak when exposed to moisture prior to sealing, and the bag could be opened and closed without any evidence of tampering. That had to be an inside job since those who packed and closed the bag had to communicate with whoever was stealing from it later. We solved that problem very easily by printing an invisible ink on the adhesive surface that smeared and changed color if licked. Any such packages would be rejected and repacked before leaving the building.

Tamper-evident technology became one of new core competencies of our company. Besides addressing the ongoing issues with the product, we also looked at other areas in which we could use our tamper-evident technology, and we expanded into forensic-evidence bags and other packaging areas. The new acquisition resulted in several inventions and patents for our company.

Inventing through Product Extensions

You can often end up in new inventions through product extensions. When you have a product that has been commercialized successfully, you looking for ways to increase your sales, profit, and market share by finding other uses for the product or modify it to use it differently.

After Bubble Wrap became a huge success, the R&D team got busy trying to expand the business to other areas. The first attempt was to try this for a nonpackaging application. Before I joined the company, the R&D team came up with a brilliant idea to use the Bubble Wrap as a cover for swimming pools. The idea was to place a sheet of Bubble Wrap to cover a pool while not in use. This pool cover idea was suggested for three reasons—to heat the pool by letting the sun in, to prevent heat loss by having the bubble insulation on the top, and to prevent water evaporation.

But no invention is simple. The concept was good, but the challenges were many. You have to evaluate every aspect and every detail before you can finalize your proposal. For many reasons, one product used for certain application may not be suitable for other applications. In this particular case, it was the difference in environment in which the product was used. When used as a protective cushion, Bubble Wrap is never exposed to sunlight. However, if it is exposed to sunlight for a long time, the polymer material the bubble is made from tends to degrade and disintegrate. After extended use, the product will fall apart in the pool and make the customers very unhappy.

We identified that problem early on through the normal evaluation process, and the R&D team started to work on the problem. They identified UV-absorbent chemicals that could be mixed with the polymer before making the bubble. The bubble shape was also modified to minimize the air trapped in each bubble to reduce the volume during shipping.

The swimming pool cover became another successful and very profitable product. The company enjoyed this success for many years until it came to an abrupt halt. The company took all safety precautions—it informed customers about the potential hazard if the product was not used properly. Warnings were printed clearly on the product instructing the user to remove the cover completely from the pool before entering the pool. We knew people could suffocate if they got under the cover. In spite of this common knowledge and the clear warning labels, there were a few incidents in which children were caught under the pool cover, and we were sued. The company eventually took the bold and responsible decision to discontinue this product.

It is extremely important to look beyond your invention and evaluate all the safety features. Product safety and liability are very critical factors in successful product commercialization.

Modifying a Product for Other Applications

Another example of a Bubble Wrap product extension was the bubble-cushion mailer. People would wrap small objects in Bubble Wrap prior to putting them in envelopes of various sizes. Someone in my company had the vision to create a cushion mailer using Bubble Wrap.

This product evolved. While wrapping a product with a bubble sheet, somebody thought it would be faster to pack the product if the bubble sheet was shaped like a pouch, and soon the first bubble pouch was introduced making packaging much easier.

The invention did not stop there. The first cushion mailer was introduced by laminating a paper film to the outer surface of this pouch. The paper provided opacity to the pouch to hide its contents. The paper also provided a smooth surface to write the address and to place stamps on for mailing.

Years later, plastic cushion mailers were introduced by replacing the outer paper with a plastic film. For this product, a paper-like plastic film was developed that could be written on and accept gum labels like paper. Replacing paper with plastic has two major benefits. The outer plastic film provided a waterproof mailer and provided better protection for whatever was inside. The most important benefit was the fact that the mailer was made from plastic and could be recycled.

Customers are always looking for things that are better, faster, and easier to use. The R&D staff will always be busy trying to keep up with customers' demands. In this fast-moving world, when you look at a product today and compare it with the original product, you probably won't believe the transformation. Some of the other examples of Bubble Wrap product extensions were anti-static bubbles for packing electronic equipment, flame-retardant bubbles for fireproof packaging, multicolor bubbles for special occasions, and laminated bubble products for various specialty applications including insulation.

Adding New Features for Other Applications

Many new proposals were made to modify the Bubble Wrap product for other applications. One such application was to make a fragrant bubble with a matching color. Not knowing the market potential, I decided to make bubble samples with certain fragrances and matching colors. Sometimes, you believe in an idea that you come up with and get excited about it, but you are not sure how the sales and marketing teams will react. I took this on as a fun project and a personal challenge. I did not solicit support from sales and marketing; I worked on this project in my spare time with my boss's full support.

My thought was that a fragrance locked in a polymer matrix would release slowly and would thus continue to provide the fragrance for a long period as the inner surface of the bubble retained and locked in the fragrance. A fresh burst of fragrance would be released each time a bubble was popped possibly even after years of storage. Two fragrances that I tested were strawberry and pine scents.

After identifying the right fragrances and colors, I reformulated the polymer recipe with these ingredients. Since I did not have much machine time available for such projects, I made sure I did my homework to get the right formula and the machine-running conditions ready. The first day, I made the bubbles with strawberry fragrance and color; the fragrance filled the plant, and everybody loved it—it was a big hit by the end of the morning. But by evening, the fragrance was too much for the plant people to handle, and they started to complain.

Other than the odor complaint inside the building, the color and the fragrance for the product came out perfect. The next day, I repeated the test with pine scent and green color and made a good product.

The week I retired, I happened to test the fragrant bubble sample I had made twenty-five years earlier. I was surprised that I could still smell the fragrance on that bubble sheet.

Missing Opportunities by Giving Up Too Early

We had created colorful bubbles by incorporating color chemicals into polymers during the plastic-processing step. The process had limitations since it could produce only one color at a time. The iridescent plastic film that was available on the market came in multiple shiny colors, and that intrigued me. I wanted to create a similar color display feature for bubbles. After finding out the technology behind creating that kind of color feature, I realized it would be a very difficult process that would require costly equipment. There was no way I could justify such a project.

But instead of completely giving up on that project, I kept on thinking of other ways to create a similar effect on the bubble. One day, I manually tried to convert two iridescent films into a bubble shape. One of the films was thermoformed into a bubble shape and sealed to the second flat film. The result was disappointing; it was difficult to thermoform—that is, to create a bubble shape by applying heat and vacuum—the iridescent film without destroying the color features. Even if it were successful, it would be a very costly process and thus a very costly product.

I thought I had tried everything, but I was not about to give up. One day, I was working on an entirely different Bubble Wrap project. I cut out two pieces of bubble sheet to compare the properties. While examining one bubble sheet, I placed the other bubble sheet on the table. After I finished with the first bubble sheet, I turned around for the second sheet. I realized there was a piece of iridescent film on the table and I had actually placed the bubble sheet on that film. I could not believe what I saw. The Bubble Wrap had become iridescent just by placing it on a sheet of iridescent film. I realized that the iridescent film effect was reflected through the entire clear bubble film when they were placed side by side. I dropped what I was doing and sealed iridescent film to the back of some Bubble Wrap. I made a perfect iridescent bubble without any costly equipment or process.

It's common to become frustrated when all your attempts fail and you think there's nothing else to try. But rather than giving up, put that thought in the back of your mind and take a break. You're bound to get a clue somehow, and the answer may pop up right in front of you.

Taking Advantage of the Other Features in a Product

When you look at a product, you may see some additional features that were not intended for the proposed use. Many have taken advantage of these features to create new products and possibly new inventions. Years after Bubble Wrap was invented, many ideas were proposed for other potential uses for this product. Some of these ideas came from customers, and others came from within the company. One such idea resulted in a very successful product—someone noticed that the air trapped in each bubble could provide good thermal insulation.

Working on this idea, they found out that since the bubbles were made of polymers that melted at low temperatures, so they couldn't be used in high temperatures as insulation. But then someone suggested laminating thin aluminum foil to the bubble sheet. That was a brilliant move that besides solving the problem created a few additional valuable features. The foil protected the product from higher temperatures and provided some flame-retardant properties and best of all created a new reflective insulation property. Besides providing conductive insulation by the air-filled bubbles, the laminated foil also helped reflect heat away from the product and made it a superior insulation product. The result was the creation of a value-added superior product from a commodity product like Bubble Wrap.

And this idea was extended even further. Using the foil-insulated bubble, we created a new insulation bag for shipping temperature-sensitive products such as medicine and food. That was a good solution for an ongoing problem in the shipping industry. Using different-colored foils, the foil bag idea was extended further to other areas like decorative pocketbooks and gift bags.

People with creative minds will always keep their eyes open and are likely to see things others don't. Hundreds of uses and applications have been reported for Bubble Wrap. Take the challenge and see how many uses you can come up with!

Improvise to Invent

I happened to have the hottest office in our building during summers; it was a corner office with windows on two sides. In spite of the air conditioning, the summer sun hitting on both windows made the room uncomfortably hot during peak hours. Since I sat in front of my computer most of the time, I thought it would be nice to have a fan on the top of the computer screen blowing on my face. I took it as a challenge and decide to come up with a creative solution. Since I got plenty of sun in my office, I decided to build a solar fan.

I got a small solar panel, a small fan, and wire to connect them. I made a cone out of cardboard and placed the fan at the smaller end to get a good airflow through the larger end. I put the conical structure on top of my screen with the fan facing me. I attached the solar panel to the window area where I got the most sun. The entire project took me about an hour, and I enjoyed nice breezes every time I was in front of the computer. I felt good about this project especially due to its environmentally friendly approach. I could have bought a fan that I simply plugged into an outlet, but I created a sustainable product without using the building's electricity.

Of course, solar fans are not new, but what is important here is how I improvised a solution. I got many compliments from my coworkers. When you need something, accept the challenge, be creative, and figure out how to get it.

One Idea Leads to Another

Once I came up with my solar fan, I thought there might have been other similar areas I could explore. The first thing I thought about was outdoor applications—military personnel and construction workers wearing helmets out in the sun. I guessed that they would like a solar fan inside their helmets. I drilled a hole into the top of a hard hat large enough to fit a small fan. I attached two small solar panels on both sides of the hard hat and connected wires to the fan. It happened to be a hot summer day, so I went outside wearing my new hat. The fan worked well; in spite of the hot sun, I felt a cool breeze inside my hat, and I smiled. A few of my coworkers who were smoking in the parking lot saw me standing in the sun and smiling. It took me a while to convince them that I wasn't crazy!

When you come up with an idea, work on it to make it reality. If you ponder that idea long enough, you might discover other useful solutions. Chances are that some of the new ideas are even better than the original. This happens more often when you really focus on a problem.

Making Use of Features for Unrelated Applications

Many years ago, one of our customers came to us for a bubble product that would dissolve in water. They were looking for a pouch containing certain chemicals that could float and dissolve slowly so it would gradually sink to the bottom and distribute certain chemicals evenly in water. The initial thought was to make a pouch with water-soluble polymers. This idea did not work, as the entire pouch sank to the bottom as soon as a portion of the bag was dissolved letting water replace the air in the bag. They wanted a pouch that would float even after partially dissolving.

We tested a bubble pouch made from water-soluble polymers. We put it in water and saw that the pouch floated well above the waterline as all the bubbles were filled with air. The outer film on each bubble started to dissolve very slowly and release the air inside. As more bubbles dissolved, the pouch started to slowly sink and release the chemicals on its way down. We came up with a bubble-film ratio to match the required rate of chemical distribution.

That was an example of how one product can be transformed into a product with a different application. This new application was never intended for Bubble Wrap, and I'm sure the original inventor never thought of it that way. Look around and you will see many examples of an invention having been transformed into unrelated products and handling different applications.

Creating a Product Is Never the End of an Invention—It's Only the Beginning

When Al Fielding and Mark Chavan, the inventors of Bubble Wrap, invented the original bubble machine in a small garage in the 1950s, their idea was to use this product as fancy wallpaper. A few years went by without any success in penetrating the wallpaper market; they thought their idea was a total failure. But then someone suggested that their invention could cushion products during packaging and shipping. The rest is history.

Coming up with a product is only a part of a good invention; identifying its full potential is equally important. Bubble Wrap was invented as a wallpaper, but instead, it became swimming pool covers, bubble-cushioned mailers, foil-laminated reflective bubble insulation, inflatable bubbles, et cetera.

Cost Factors in an Invention

Regardless of how good a product is, people will be reluctant to buy it if the price is too high, so costs need to be taken into consideration in every step of the invention process. There are ways to justify a higher cost. One approach is make the product more appealing by creating value-added features such as better performance, ease of use, longevity, compactness, reusability, recyclability, or degradability.

At one point, I worked on a sustainable paper cushion product to make it a very low-cost product and a machine with which to make it. I decided the basic component to be a low-weight, crude brown paper, which was the lowest-cost raw material available at that time. When I started, I had no idea what the final product would look like.

I tried to create many types of paper shapes by folding, twisting and crumpling. I was intrigued by one particular shape—a rope created by twisting the paper. I thought I was on to something. The twisted paper with air trapped inside gave me a bulky cushioning product. A piece of thin paper was turned into a three-dimensional structure with pretty good cushioning strength. The next challenge was to find a process to make it. I pulled the paper straight off the roll, and it came out as a flat sheet of course, but when I pulled it from an angle, the paper curled as each layer was unwound; I was getting the tubular shape I wanted without any external twisting action.

The paper roll had to be suspended in air for the paper to be pulled off the roll easily, but I knew there had to be a better way of doing this. I tried pulling the paper from inside the roll and discovered I could easily do that and create a continuous twisted tube without suspending the roll.

The twisted paper product provided some level of cushion protection, but it tended to unwind. I tested various methods to keep the paper from unwinding including crimping and using glue. They worked, but that of course added cost to the process and thus the product. I had pretty much run out of ideas to improve the twisted paper by that point because I was too focused on using a single twisted paper. Thinking outside the box, I came up with the idea of braiding two of these twisted lengths of paper. The braiding action locked the twisted paper tubes together and provided superior cushioning properties.

The challenge then was to find the most efficient way of braiding the paper. Two twisted paper tubes were delivered side by side by pulling paper from within the cores of the paper tubes. Braiding these twisted tubes by pulling them in a continuous fashion was very difficult for a continuous throughput. Instead of braiding at the receiving end, I decided to see if I could braid them at the delivery end. The paper rolls were placed in containers attached to the opposite ends of a panel. The center of this panel was attached to a spinning mechanism powered by a simple electric drill. The two paper rolls could rotate around their centers. While spinning, a braided paper tube was created by pulling forward both the inner layers of paper together. The braided product could be wound on a roll or cut to length without unwinding the braided rope.

I made the first prototype machine that could produce 200 feet of paper rope per minute. This prototype machine cost under $25! Low product cost is very important, especially when you are dealing with commodity products.

Invention to Make Life Easier

If you drive to work every day, chances are that the sun gets in your eyes when you're going to or coming home from work. You try pulling the sun visor down to the lowest position or sitting up higher. After many years of going through this frustration, I thought there must be a solution to this problem.

My goal was to create a sun blocker that could be moved around, up, down, and sideways to block the sun from any direction. I created a movable frame attached to a vertical rod. The frame was able to move up or down and fix at any position. The vertical rod was attached to a collar. A horizontal curved rod was inserted into the collar so the collar could move side to side. I designed a hook mechanism to attach the curved rod to any sun visor and then be removed and used in any other car. I made a prototype of this adjustable sun visor and tested it.

The new design worked very well. The sun-blocking frame was easily adjustable and could slide down, especially when the sun was closer to the horizon in the mornings and evenings. Drivers could move the frame to their right or left if that's where the sun was coming from. The curved rod actually brought the frame forward on both sides. I also modified the blocking frame so that any semitransparent color panel could be inserted in the frame to provide a sunglasses effect.

I got negative feedback when I presented this idea to our company. The response was that, being an automotive part, it was not something our company could market as we weren't in the automotive market. Even if we were to market it, the company thought it would be a very high- liability product with unnecessary risk.

Inventing a product is not easy, and successfully marketing it is an even bigger challenge. Before you select an area in which to invent, focus on what is in line with the manufacturing and marketing capabilities of a company. You will have a greater chance of success if you avoid areas of high risk and liability.

Addressing Customer Requests

During the early years with this company, we had a small division that made paper bags for takeout food. Our customers would use staples to seal the bags, but they were getting many complaints from customers about the metal staples getting into the food when they opened the bags. They asked for our help.

I listed possible solutions including adhesive tape, an adhesive-coated flap on the bag, rubber bands, mechanical twist ties attached to the bag flap—my list went on. The customers were very picky and did not want any changes to their normal operations. They also said it would be nice if they could avoid adding any foreign objects to the bag. That made the challenge greater. I made prototypes of the selected ideas from the above list. Some of them worked very well, but I had to use a foreign object such as tape, adhesive, or plastic twist ties. The customer was not fully satisfied and wanted us to come up with some more ideas.

With the requirement that there be no foreign objects, I tried folding and creasing the bag different ways to see if it would stay closed. Some of the multiple folds performed better than others did, but none was satisfactory. But then, I folded the flap the normal way and tore a portion of the flap in the shape of an arrow. I pulled the arrow-shaped portion back and folded it in the opposite direction. I noticed the second fold on the torn piece locked the first fold and did not open easily. The arrow shape on the torn piece was stuck in the narrow gap created during the initial tear. This actually acted like a staple attaching paper.

It was a good idea, but I could not ask the customers to tear the paper bag in a particular way and fold it before they sent it out. I knew I was on to something good, but I didn't know how to proceed. I put on my mechanical engineering hat and started to figure out how to automate this tearing process. The thought of a paper punch came to mind. I thought that if I could change the shape of a basic paper hole punch from a circular die to an arrow-shaped die, I would have exactly what I was looking for. A paper punch cuts out a hole completely, but the punch I wanted would make an incomplete arrow-shaped cut; the bottom of the arrow would still be attached to the paper.

I was halfway there. The challenge then was to pull the arrow-shaped portion back through the slot to lock the papers. I modified a paper punch die and hole to create the arrow shape I was looking for. It was a crude prototype, but I had proven my theory with it. When I pulled the tool back after punching, I noticed that the narrow paper strip folded by itself to lock the papers together. I was surprised at the fact that the arrow-shaped strip was caught on the die lips, which pulled it back to fold backward. That was exactly what I wanted. That's what I call luck!

THE INVENTOR IN YOU

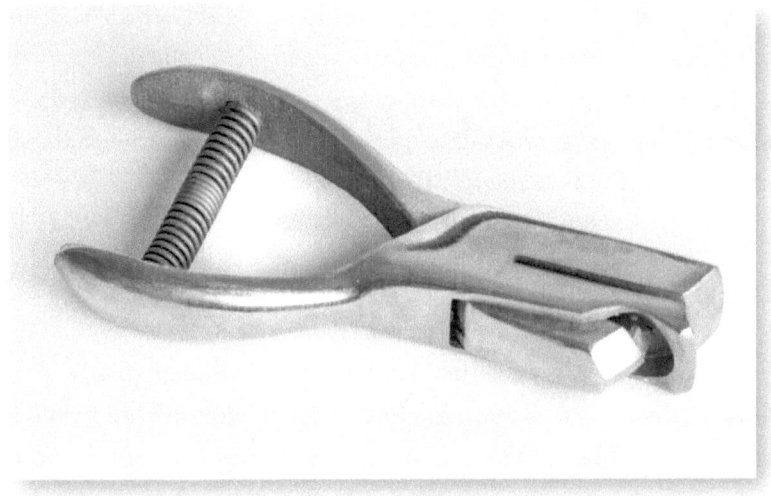

Stapler without Staples

I took my drawing of this design and showed it to a tool and die maker. The maker was able to make the samples for us in less than a week and at a very affordable price. And our customer was very happy that we had given them everything they had asked for. It meant no more complaints about metal staples from their consumers, no more need to buy and stock metal staples, and no foreign materials used to fasten bags.

Giving customers what they need is critical for sustaining and growing a business. When you face challenges and failures, keep trying and explore all possible avenues. Don't give up, because sometimes a little luck will come your way!

Special Applications

Our team worked on a special bubble product for Halloween. We wanted to make Bubble Wrap that glowed in the dark. We needed to add chemicals that glowed in the dark to the formula for Bubble Wrap and did just that; children in costumes made of this would be safer at night, and we thought it could also be used in Halloween decorations.

But because it was such a seasonal product, it did not fare well in the market.

One of the problems with entertaining special requests from customers is that in most cases, it will be a low-volume market and thus will be difficult to justify the R&D efforts.

Adapting New Technology

While on vacation in China, I saw a toy—small, flat, plastic pouches shaped like animals. When struck, the flat pouch inflated into a large plastic animal. I bought several of them and brought them back to my office. Out of curiosity, I opened one and saw it had another small plastic pouch filled with a liquid and some white powder around it. When the liquid pouch was broken, its contents came in contact with the powder, and their reaction released gas that filled the outer pouch. I immediately identified the liquid as vinegar and the powder as baking soda and was able to reproduce the expanding pouch.

I decided to apply this principle to Bubble Wrap to make an on-demand, self-expanding cushioning product. I made several samples of flat bubbles, each having a tiny sealed pouch of vinegar and a little bit of baking soda. Each bubble was inflated instantly when the internal mini pouch was broken to release the vinegar. One of my prototypes had multiple flat bubbles in a continuous film; the plastic passed between pressure rollers that broke the vinegar pouches so the vinegar and baking soda could react; the bubbles expanded as soon as the film came out through the rollers that had crushed the vinegar packet.

I wanted a product that could be shipped flat to save shipping costs and storage space; the customers could expand it on demand and create only as much as was needed at any one time by passing it through rollers. It sounded like a good idea, but it failed miserably on the market. The main reasons were the costs of the chemicals and the labor to pack them into each bubble as well as the presence of a liquid in each bubble. Most customers didn't want foreign chemicals in the bubbles that could potentially contaminate whatever was packaged in them. We learned that early on and decided not to proceed with that idea.

Regardless, this is a good example of how an inventor can apply a technology used in one area to another area. Though this one didn't work, we at least learned how such products could be made; we had developed a process to inflate the bubbles on demand. This wasn't a solution for cushioned packaging applications, but it could help find solutions to other problems.

Process Inventions

Almost all new product inventions are patented; you won't see as many patents for process inventions. One of the main reasons is that any one company in an industry wants to protect its processes from the competition. Even though a patent will provide good protection, in certain cases, unlike a product patent, it is very difficult to police others' processes to see if they are infringing the patent. Unless there is a clear way to police and determine if others are infringing, the industry stays away from filing patents for the processes they develop; they consider them trade secrets.

Over the years, I have come up with several inventions we used in the process of making certain products, and we have kept some of these as trade secrets.

About the Company

It was a real pleasure to work for Sealed Air Corporation for thirty-four years. I enjoyed my vacations and holidays, but I always looked forward to going back to work and getting busy again. I always kept my project challenges in the back of my mind and reflected on them every chance I got or every time I saw or heard something new. You'll be surprised how many new ideas or clues you will get by making a habit of such reflection even outside work.

I had very close relationships with the first two CEOs of this company. They were very fond of my work; they read my monthly reports regularly and gave me feedback on occasion.

The first CEO, Dermot Dunphy, used to come into my office and ask, "Charlie, what have you invented this month?" He always supported me with my projects. One time, I was working on a project for which I did not have the full support of the marketing department. When he found out about my project, he told me he believed in it and would support me even if he had to twist the marketing department's arms. It is not very often that you find people like him at the top. With support like that, you give everything you have to make the company successful.

I have a funny story to share with you. I attended a sales and marketing meeting in a neighboring town. Everybody was free the afternoon on the final day of the meeting. We were invited to play golf or tennis, and since I was a good tennis player, I signed up for tennis. Dermot saw my name and told me that he would like to challenge me in a doubles game, so we sought out partners, and I showed up with one of my coworkers. To my surprise, Dermot showed up with a tennis pro. Dermot was pretty much a beginner in tennis and couldn't handle the ball well, but the pro covered for him well, and they won. At the dinner that night, Dermot made a speech to the employees in which he joked that he had had a good tennis match with two youngsters and had beaten them badly!

After Dermot retired, Bill Hickey became the new CEO. I had known Bill for a long time, and he also was very familiar with my work. Bill, an engineer, paid more attention to the R&D projects than Dermot had, and he wanted to know every detail. He actually worked with me on one of my projects, and we ended up filing an invention disclosure together for a patent application.

I once went to India on a vacation for a couple of weeks. While I was there, a tsunami hit the southern part of India. I was only ten miles from one of the areas destroyed, but I was safe. Because of the instability and confusion in the area, I was a few days late getting back. Bill knew I was vacationing in that part of India. As soon as he heard about the tsunami, he made inquiries about me. He asked the secretary to trace my flight schedule and get my travel details. Since I did not get back on the scheduled date, everybody started to worry about me. The secretary kept calling my home phone; everyone at work was concerned. All my coworkers were relieved and happy to see me back in the office. Bill personally called me as soon as he got the news.

After we acquired a large food-packaging company, we created a new division to handle its products. Bill wanted me to get involved in the R&D for the new division, and I did. It was a refreshing and exciting change. Compared to protective packaging, food packaging had high standards and strict regulations. Eventually, I was moved to corporate R&D, where we had the R&D responsibilities for all the company's divisions. I was very fortunate in that I broadened my experience by working in various divisions. I am proud to have come up with many inventions and received patents for products and processes in many product lines in all the company's divisions.

I want to thank both Dermot Dunphy and Bill Hickey for all their support and for appreciating my contribution to the company.

Sealed Air Corporation always supported and promoted inventors and recognized their valuable contributions. The company inducted any inventor with twenty or more patents into its Inventors Hall of Fame, and I was inducted in 2004 with twenty-four patents at the time. I retired from this company in 2017 after receiving sixty-nine patents—thirty-five in the United States and thirty-four internationally.

After learning about my inventions, the Indian community in the New York area honored me with the Engineer of the Year Award in 2010. I have eleven more patents pending. Over the years, I have come up with over three hundred ideas for inventions that are all well documented in Disclosures of Inventions filings. I am still working on some of these ideas unrelated to my previous employment. I was still busy filing two invention disclosures on the day of my retirement from Sealed Air Corporation.

CHAPTER 4
INVENTION AND PRODUCT DEVELOPMENT

The word *invention* has many definitions. US patent law defines an invention as "a new, useful process, machine, improvement, etc., that did not exist previously and that is recognized as the product of some unique intuition or genius, as distinguished from ordinary mechanical skill or craftsmanship."

Many ideas can lead to an invention. An invention can be the result of someone's attempt to solve a problem or meet a specific need. The need may result in a new product or process, improving the properties or appearance of a product, reducing its costs, adding new features, making the product environmentally friendly, improving the steps in a process—on and on.

And modifying an invention or finding a new application for it could lead to a new invention. Invention can also be the result of an accident, mistake, or a test producing the opposite or unexpected results. There is no way of telling how long one needs to work to complete an invention; it could take a very long time or could be spontaneous.

Market-Driven Inventions

Most inventions occur when one tries to develop a product or a process independently or at the request of a customer. The sales and marketing teams in a company can decide to fill a void with a new product, compete with existing products, or enter a new market. In all these cases, the product needs to be unique and competitive and can be protected by intellectual property rights. After their evaluation, sales and marketing define the product concept and present it to the development team. Any projects initiated this way are identified as market-driven projects. In this scenario, we are mainly dealing with product ideas.

In an industry setup, once an idea or concept is defined and approved, it needs to be developed to meet all the requirements. At that time, the developing team becomes part of a bigger team that includes sales and marketing, supply chain, finance, and legal.

Once a development program is in place, the team goes through a brainstorming session during which everybody on the team contributes ideas and works toward the common goal of developing that product. During the brainstorming session, an important rule is that no idea is considered bad and all ideas will be accepted for further consideration. This will give every person a chance to think freely outside the box. I was in countless situations where certain ideas that were considered strange and branded as unworkable at first turned out to be real solutions.

There are many ways to develop a product to compete in the market. The easiest way is to reverse engineer an existing product to come up with a copy; that's considered okay if the existing product is not intellectual property. You may end up adding new and valuable features to the product during your reverse-engineering process that could lead to an invention that could be considered intellectual property. Though you cannot patent an existing and protected product, the modifications you make could be patentable. If you succeed in getting the patent protection, you can enjoy the entire market share for this modified product without fear of competition.

If an existing product is well protected by intellectual property rights, you won't be able to infringe those rights. But it's important to study the existing patent because not all patented products are well protected. Some patents may be weak with poor and narrow or limited claims leaving a lot of room for competitors to work around the patent and come up with something that doesn't infringe the rights of the existing patent holder. That has happened frequently. I've seen some cases where the competition worked around a patent and made a product even better and then patented it. That put them in a position to succeed in the market for that product.

Most countries respect international patent laws, but copying and marketing patented products happens in certain parts of the world where patent laws are not strictly enforced.

How can you compete with a patent-protected product? Besides working around the product and modifying it, you could develop a different product that meets the same market requirements but is entirely different from the one you are competing with. These new products may entirely replace the competing products. Some radical and well-known examples of this include replacing vinyl records with cassette tapes that were replaced by digital music, replacing VHS videocassette tapes with DVDs, replacing photographic film and cameras with digital cameras, replacing landline telephones with cell phones—on and on.

The younger generation may not remember or have any knowledge of older methods of communication. Telegrams were the oldest form of modern communication, but they were replaced by telexes, and then fax machines, emails, and now text messages. These newer products did everything better and faster than the previous products did and had many more features that improved customers' lives and experiences. The idea is to create a product that appeals to customers by making it more efficient, cheaper, more durable, longer lasting, more compact, and easier to use. The more you can do this, the more valuable your invention will be.

And with so many alternative approaches with which to compete in a current market, you will have a better chance to secure a strong patent with broad coverage. Of course, your success depends on how close the new product comes to the above ideal situation.

The aforementioned inventions definitely had a huge impact on the market, and they changed the way we do things. I am sure nobody expected such huge successes when they were working on these inventions, which were not overnight successes; rather, they became successes gradually. The most important fact was that these inventors had a vision for change. They would have been taking a big gamble if their vision and focus had been aimed primarily at achieving overnight success. For example, before the cell phone era, the idea of a portable phone replacing all landline phones would have been far-fetched based on the existing technological and marketing challenges.

In this world of fast-moving technology, it is important to take a proactive approach, think ahead, and prepare for the future. If you don't, you'll miss opportunities and your product will run the risk of becoming obsolete. You might find it necessary to develop new products to replace your old products not necessarily to increase your market share but simply to keep it. It is better to cannibalize your existing product rather than lose business to competition. Look what happened to Kodak, a giant company, and how it disappeared. This film and camera manufacturer did not foresee technological changes and was thus not prepared for them. Its products were replaced by digital cameras, and it was forced out of business.

In any industry, developing a product is only part of the equation; innovation is an important factor as well. Ask the following questions before moving forward with an idea for a product: Is this product unique enough for customers to become excited about it? Will customers embrace it or be reluctant to switch to a new product? Can this product be manufactured economically? Does it meet all legal and industry regulations? Can it be marketed at a competitive price? Can it be supplied in volume in a timely manner? But the answers to these questions do not matter if your product is not protected by intellectual property. You need to have all the right answers for successful product development and commercialization.

If you come up with a product that looks, feels, and performs like a competing product, customers won't buy it unless you can sell it for less. Customers are always looking for features including better performance, a more attractive appearance, longer product life span, ease of use, lower cost, and so on.

Creating a value-added product is always the best approach for product development. The more value you can claim for a product, the better customers will receive it. In industry, most customers are tough negotiators, so adding value to a product always helps to win them over. Some customers are even tougher as they want all the new features but aren't willing to pay more.

The success of an invention will also depend on how easily it can be manufactured and marketed; you might not be able to manufacture it successfully. You may run into different problems, issues, and roadblocks trying to develop your idea into an actual product.

Once you have a general idea about your new product, the first thing you need to do is make a prototype of your concept. You will begin to see these aforementioned issues as you make your prototype, and that will help you identify and address the issues early on.

Once you have the working prototype, the next step is to see if it can be manufactured efficiently and economically. The supply chain will determine if this product or process is supported by your company's core competency. It will be a concern if it requires an entirely new process or machinery that the company isn't familiar with. As an inventor, your responsibility is to take your company's core competency into consideration before you think of inventing for the company. You also want to make sure that all the raw materials required to make this product are readily available and in ample supply.

The new product and process may also require government regulatory certifications from agencies such as the FDA, EPA, and others. You need to understand the new product and process thoroughly to determine what certifications are required, and those certifications may take a long time to get and could be very costly. The supply chain needs to be comfortable with the product and process before you decide on the type of products your company can invent.

As I mentioned, when our company was strictly a protective-packaging product manufacturer, I came up with a product for a medical application, but we discovered that we couldn't market it to an industry we had never had contact with. Our potential customers wanted products made by companies that specialized in medical products, which we did not.

Cost is also an important factor for successful commercialization. An expensive product can easily price itself out of the market if the cost does not justify the need. The key is to identify customers' real needs and meet them in the most cost-effective way. You can also create additional values that might provide unexpected benefits to your customers. When customers realize they will enjoy savings down the road, they may be willing to pay more for a product.

Your success will depend on how good of a marketing team you have in your company. Even if you come up with a superior product at a great price and know you'll have many customers for it, your success won't last long if you fail to protect your intellectual property rights for it. Without protection, it will be only a matter of time before the competition will start making the same product and compete against you. And your competitors will have a definite advantage in that case because they have not borne any of the development costs. It could become even worse if the competition improves your product and patents those improvements. That could prevent you from competing in the market for the modified product. It is foolish not to protect your invention before trying to market it either by patents or making them trade secrets. More on that later.

Research-Driven Inventions

Independent inventors have more flexibility; they can invent first and worry about innovations later. That may be okay as many innovators can commercialize an invention successfully. Inventors can sell their patents or negotiate a royalty or both. The work is much easier for inventors in this category since they focus only on the invention part. They have fewer restrictions during the invention process as they are less concerned about the cost, manufacturing, marketing, and industry regulations.

Research-driven ideas are those that a development team or an individual working for a company comes up with. They have to sell the idea to the sales and marketing teams, which is probably a tougher route to take since the idea has to undergo market analysis. Usually, such proposals end up getting a lower priority than do market-driven proposals.

In some cases, inventions are process related, so the marketing department has a smaller stake in the decision making. The supply chain will have more control over such process development. Normally, we see both product- and process-driven ideas in this category.

One reason research-driven ideas get low priority compared to market-driven ideas is that finding a home for that idea requires additional steps including market analysis and market research. You may invent a new product with a certain application in mind though you probably won't find out if it will be successful until you try.

When you start thinking about developing a product, it is important to conduct a feasibility study for the manufacturing and marketing capability. For example, the new product needs to be compatible with the existing manufacturing process to minimize any manufacturing challenges. In the same way, the marketing department needs to be comfortable with the new product so it can handle it through existing marketing channels.

You can come up with any number of new products, but unless they can be manufactured economically or have marketable features, they will sit on the shelf forever. Smart inventors always look beyond inventing the product; they don't want to invent something for the sake of inventing; they focus on inventing things that can be successful whether they are new products or modifications of existing products.

HOW DO WE INVENT THINGS?

CHAPTER 5

I have outlined three steps in the invention process—why invent, what to invent, and how to invent. A systematic process of invention can be described by answering these three questions.

Why Invent?

You don't want to invent something without a purpose, so the first thing you do is lay this foundation by identifying the purpose, reason, or unmet need for your invention; you need to understand why it is necessary. In this step, you identify a problem or recognize an unmet need. This is the trigger point for most inventions, and it serves as a platform for the inventor to work on.

Basically, you have been made aware of a problem that needs a solution or a need that needs to be fulfilled. Imagine someone getting a flat tire while driving on a rough road. It takes a long time for that person to get the tire fixed, and it makes him miss a very important appointment. In frustration, he yells, "I wish they made a tire that wouldn't go flat!" That makes you recognize an unmet need—a tire that can withstand rough roads.

What to Invent?

Once you have identified the problem and recognized an unmet need, the next step is to define the goal and identify various options to meet that challenge. The problem of the tire could be solved by coming up with a puncture-proof tire or one that replaces air with something else when it is punctured. But coming up with possible solutions does not constitute an invention. At that point, you have only some basic ideas you are not sure will work. But you need such ideas for the next stage of development that can lead to an invention.

How to Invent?

In the above example, you have come up with a few possible options to meet that challenge. In this stage, you will determine how to develop these ideas into a working solution. You will need to examine each option and learn what is involved in the development of each option. Usually, you select the best option to work on, but multiple options can be worked on in parallel. To reach this goal, you will have to go through a series of research and development processes. In this final stage, you generally end up with an invention for a product, a process, or both. When you have multiple ideas, you can prioritize them by comparing each one's ease of development, available technology, cost of development, and of course the possibility of protecting the intellectual property rights.

Inventors who wanted a puncture-proof tire might first look into redesigning the rubber formula to make it tougher, or by coating the outer surface with tough, durable materials, or laminating the rubber to a metal plate. Inventors would look at the second option in a similar way.

Here again, there are different ways to approach the challenge of replacing air with something durable so the tire remains full even with a puncture. Inventors could develop a foam flexible enough to fill the tire instead of air or a series of springs that provide the same flexibility as a pressurized tire. In these cases, the tire would remain full and wouldn't go flat even if punctured.

At the end of this analysis, the inventor would have a basic idea about which of these options would get high priority to move forward with the development process. This final research and development process brings it to fruition by creating a proven model that meets the goal. This generally results in an invention for a product, a process, or both. The real work of an inventor is in the second and last steps, and these are the most important and difficult factors of an invention. Let us look at these three steps in detail.

The solution to the above example may look complicated and would require intense research and development. I have given this example on purpose to show a systematic approach to an invention process. This does not mean that all inventions are this difficult and all inventions have to go through such a detailed process; we will look at simple examples at the end of this chapter.

An inventor may not necessarily get involved with the first step in which a need or problem is determined. Often, that's done by somebody other than the inventor. This is especially true if you are working for an industry. For example, in many cases, a customer may have a problem with your product that needs to be solved, or a request for a new product to meet a need, or the desire for you to modify your product to meet certain additional needs. Market surveys by the sales and marketing group may result in new ideas to expand your product profile and business. Independent inventors, who are not associated with an industry, usually come up with their own reasons for an invention.

When you identify a problem and recognize an unmet need, the goal may be to create certain new things or modify an existing product. Once you have identified a new problem to solve or are looking for a better answer to an existing solution, you first need to thoroughly understand the challenge. You must analyze each of the current issues to figure out exactly what a customer is looking for, and you need to understand the problem better than the customer does to create the right solution. Sometimes, the customer may not have the proper insight into the problem and will tell you what they want, but the customer's perception of the solution may not be the right one. What they say they want may not be what they need. An inventor can analyze the situation better and can offer what customers need rather than what they want.

Independent inventors may need to create their own reason to invent by keeping their eyes and ears open all the time looking for opportunities. Any time they see or use something, they ask questions:

What else can I use this for? How else can I use this? What about it do I like or don't like? How can I improve it?

Inventors generally are not satisfied with what they have, and they always want things different or better, so they challenge the status quo. For example, when they use a vacuum cleaner, they might think it is too heavy, too noisy, or not designed ergonomically. They might want a nozzle that can get all the way into corners and other areas that are difficult to reach. They might wish they had a more efficient way of replacing filters and so on. They have just come up with several reasons to invent just by asking such questions.

You see and use hundreds of tools and many kinds of equipment and appliances every day. With most of them, you'll probably see something you aren't happy with and would like to see it improved or made better. Try to be critical of everything—instead of seeing only the good in everything, look at the negative and identify the areas that can be improved. Just from these alone, you can come up with plenty of reasons to invent.

Inventors need critical and curious minds that are always on alert. Start asking questions—Why do this or that? Why not do it that way? What happens if I do this?—every time you see or hear something new. Always try to have an open mind; avoid tunnel vision and think outside the box. You can let your imagination go wild, and during this step, you can come up with several ideas that can possibly lead to an invention.

Another way to be creative is to combine multiple tools to do something different, better, or more efficiently. Besides focusing on modifying one thing, you can combine two things to create a single item to provide multiple functions. Let us say you just finished vacuuming the floor. The next job you tackle is mopping the floor with a wet sponge and maybe applying wax. Anybody who pays attention will probably think that maintaining floors takes too much work and requires too many steps. Now you have identified a reason to come up with something better to make your job easier.

When curious people analyze the tools used for two different operations, they will probably think there must be a better way to handle both. They get excited about a challenge; it gets their adrenaline flowing. They look at a vacuum cleaner to see if there is anything they could do to make vacuuming faster. They do the same with a mop, and then they start thinking of ways of mopping and vacuuming simultaneously—maybe one piece of equipment that has two panels, one for vacuuming and another for mopping and waxing at the same time.

I mentioned Sam Houghton, who at age five came up with a way for his father to rake just one time and simultaneously collect larger leaves and twigs and smaller bits of lawn debris, something that had required him to use two rakes. Well, Sam combined the rakes on one handle in a way that made it easy for his father to flip it around depending on the task. This young inventor was granted a patent for this invention.

There are no limits to what you can do to initiate an invention process. Another way to come up with ideas for inventions is listening to anyone's wish list, you know, the "I wish I had …" or "I wish I could …" Every time you hear that, somebody has just given you a reason to work on something new. Write down everything you hear, or make your own wish list. This will be a great platform for you to generate new ideas.

Start thinking about everything you do and use from the time you get up in the morning till you go to bed. Think about every action in detail—getting up, using the toilet, brushing your teeth, exercising, cooking, eating, going to work, driving, doing the things you do at work, shopping, playing games, outdoor activities, yard work, fixing things, using tools, relaxing, watching TV, using the computer, reading, and sleeping. Analyze each of your actions and things you use and ask yourself, *Am I happy with how I do things or with the things I use? Can I do things differently to make it better, easier, faster, and more comfortable?* I am sure you can come up with a long list of things and processes you would like to see change for the better.

The things around you have changed over the years, and they will continue to change because of the demand for new and better things. Take a simple example—the screwdriver. The first one had a simple design with a flat head that fit into a slot in the screw. Soon, the Phillips head appeared with better torque, followed by other driving heads like the Allen wrench and socket wrench. Then different sizes of driving heads showed up for handling various-sized screws.

Fed up with having to carry so many different screwdrivers, somebody decided to make a compact screwdriver set with one handle and multiple driving heads that could be snapped on and off the same handle. A simple screwdriver design led to many inventions complementing the basic idea and each time making the product better, more efficient, and easier to use. Every time you see a product, think about what the next generation of it would or could be based on what's missing in it or inefficient about it.

Product safety is a huge area for inventors to focus on. You can revisit all the items we mentioned so far and look for more opportunities in each item with safety in mind. Safety is an important factor in everything we use. We hear about so many people getting hurt from things they use every day because of faulty equipment, poor design, or improper use. Every consumer good would benefit from additional safety features to make them foolproof. As a consumer, you probably have identified at least some unsafe features in some of the things or tools you use. In spite of the many safety features that were put into this equipment, we still see people getting hurt. Sometimes, it's the result of carelessness or a disregard for the safety rules, so there is always a need to make equipment safe and then safer.

You can identify many opportunities for building safety features into these items. You can even go further to visualize an entirely different safer design or model. These could be anywhere from skid-resistant slippers or shoes to a barbecue grill with a built-in fire extinguisher. You can always add more safety features in everything you use including household items, tools, sports equipment, or any indoor and outdoor equipment. The opportunity to invent something in these areas is endless.

If you are working for an industry, there is another area you can focus on. Regardless of the type of work you do, you can contribute to your industry by doing a similar exercise. You can focus on raw materials that the company uses and products your company makes. Other areas you can focus on are material handling, the processes your company uses, its safety program, and so on. When you see and handle things every day, you become an expert with them and you start to see things others don't. You can come up with good ideas for improvement. I have seen many examples of machine operators and assembly-line workers coming up with ideas for improvement in company products, processes, and safety practices.

R&D personnel are mostly responsible for generating ideas and developing new products and processes for their industries. They have multiple sources to help them identify and recognize a need, mainly sales and marketing analyses and customer interaction and feedback. Having a good and thorough knowledge of competitors' products is extremely important. In many cases, identifying and overcoming the weaknesses in competitive products leads to inventions that benefit your business.

R&D personnel also spend a portion of their time on their own pet projects—ideas for new products or processes. Sometimes, this is done by a proactive approach for identifying issues with current products or processes ahead of time. Another approach is trying to come up with ideas for new products that complement existing products or new products that can be made using existing core technology.

Exploring new uses for an existing product can also lead to an invention with a possible increase in sales as the company where I worked for thirty-four years did. Bubble Wrap, its invention, morphed into pool covers, cushioned mailers, drainage boards for underground water control, and many other products.

Let us evaluate the second step—what to invent. Once you have identified a problem or recognized the unmet need, you think of possible solutions, and in most cases, this is done in brainstorming sessions with representatives from all facets of a company—R&D, marketing, and others. People thinking differently generate more solutions. For example, R&D may focus more on technical solutions while others focus more on other aspects based on their expertise.

Something one person finds silly might pique another person's curiosity, and that can cause a third person to come up with a good idea; in that way, ideas are built upon. At the end of a good brainstorming session, you will have multiple ideas that can help solve a problem. During brainstorming, listen to all ideas and never discount anything.

Those who work with a research team have a better chance of coming up with ideas. They have frequent brainstorming sessions for dealing with projects and constantly collaborate with each other. Any time you discuss a problem with another team member, you are sharing a lot of information related to the project. The shared information may not mean much to one person but may be significant to another. Brainstorming sessions can enrich every participant's knowledge.

Those who work independently have to come up with ideas by themselves, but they can put themselves in the shoes of someone from a theoretical R&D or marketing department; such one-person exercises can be fruitful. But they must not get hung up on one specific thing since that can hinder progress; they have to think outside the box, and that includes the box of their limitations.

Focusing only on the normal way of doing things can be dangerous. The shortest way to get from point A to point B may not necessarily be the most efficient route. You need to look at all the possibilities rather than on limited areas for solutions. The unmet need in one of the previous examples was for a faster and easier way to clean the floor. Looking only at a vacuum cleaner as a solution may generate limited ideas, but when a vacuum is mentally combined with a mop, that can broaden the scope of your ideas to create a single tool for multiple applications that effectively addresses the real issue—saving time.

Sometimes when you are alone or in a brainstorming session, you might struggle to come up with answers to your questions. But by keeping the problem in the back of your mind and thinking about it frequently, you will be surprised to find more ideas coming to you days or even just hours later.

I always used a special technique to brainstorm when I worked independently to solve a problem. Before I did any actual physical work on a project, I did some mental homework. I found that was the fastest and most convenient way to come up with ideas. I have been practicing this technique since I was very young; I guess it came to me naturally.

Every time I faced a challenge, I would go to a quiet place, close

THE INVENTOR IN YOU

my eyes, and concentrate on the challenge. I would think out loud about solutions. I would picture certain ideas, designs, or models and even manipulate them in my mind. Over the years, I trained myself to imagine how simple shapes could be converted into complicated shapes; I believe this technique is my biggest asset in that it has helped me come up with many ideas.

And over the years, I got better at this technique. Now, I don't even need to close my eyes to concentrate. I can be engaged in a conversation with a group and without being consciously aware of it, I can slip into this meditative state. Sometimes during a conversation, I hear something that I can connect to one of the problems I was working on. It may trigger an idea or a possible solution to a problem. Immediately, I go into this state of solitude, which could last for several minutes. People around me usually wonder what is going on with me. I have been asked many times if I was okay, and that can be embarrassing.

That happened to me once when I was discussing an important project with my company's president. During the discussion, which had become intense, something triggered in my mind that I thought could be a possible solution to a problem. I immediately went into my thought process by completely turning myself off from the conversation. During those few minutes, I was able to come up with a good solution to the problem. I did not even realize that the president was asking me a question. When I didn't respond, he asked me if I was feeling okay. I explained to him what had just happened and told him about the possible solution I had come up with. He was happy that I was okay and was excited about my possible solution. He joked that we should have that kind of conversation more often!

Sometimes, silly or funny statements can lead to something significant. There's an old story about a father who bought a thermos bottle for his son to take to school along with his lunch. The father explained to the son that the flask could keep hot things hot and cold thing cold. The little boy had never used a thermos before, but he was excited. He put some ice cream in the thermos as well as some hot chocolate; he thought one would stay cold and the other hot. The kid's classmates of course laughed at him, but when the boy told his father about it, the father laughed but then started thinking that this was a new need, challenge, and opportunity. He thought that if the boy used the flask for hot and cold items, there must be a need for a flask to handle multiple food products. After several months, he developed a dual flask that could maintain what was in them at different temperatures.

At the end of a brainstorming session, you'll have several possible solutions but nobody will know for sure if any of them will work. It is not possible to work on all these ideas as that would be very time consuming and costly. You'll need to narrow down the list to the top one or two ideas. You can screen the ideas by creating a table with a list of features required in the final product in the first column. You then add features that may add value to the final product at the bottom of the same column. List all the ideas in subsequent columns and start to match them against all the features in the first column. All ideas that do not have all the required features will be discarded at that point. Any idea that matches all the required features will be considered a potential "go" for the next round.

Any idea from this list that also matches value-added features would get to the top of the list depending on how many value-added features it matched. Depending on the resources, you may want to select more than one idea from the top for further development.

Chart for Screening Ideas by Matching Features

Minimum Required Features	Idea 1	Idea 2	Idea 3	Idea 4
A	Yes	Yes	Yes	Yes
B	Yes	Yes	Yes	Yes

C	No	Yes	Yes	Yes
D	No	Yes	Yes	Yes
Additional Features				
E	No	No	Yes	Yes
F	No	No	No	Yes

This example of a screening chart lists four minimum features (A, B, C, and D) required for the new product. It also lists two additional features (E and F) that can add value to this new product. The four proposed ideas were screened in this manner to qualify them for the next development stage.

This chart clearly shows that ideas 2, 3, and 4 met the required needs, but 2 did not meet any additional features, 3 matched one, and 4 matched two. Thus, 4 would be ranked as the top choice followed by idea 3 and 2; 1 would go because it did not satisfy all the minimum requirements.

Depending on funds and resources, you may want to limit the number of ideas moving forward to only one. Based on the above, you might want to quickly select idea 4, but you need to go through another screening process to rate the four ideas based on ease of development and manufacture, cost, meeting government regulations, sustainable features, and intellectual property protection among others.

In the following chart is a comparison of the winning ideas in these respects; it shows that after feasibility studies, idea 4 was no longer at the top of the list due to the problems it would present in the development, manufacturing, and cost areas. Idea 3 was then the best option to move forward although it had fewer features than idea 4.

Chart for Screening Ideas for Feasibility

Feasibility	Idea 2	Idea 3	Idea 4
Development	Good	Good	Poor
Manufacturing	Good	Good	Fair
Cost	Good	Good	Fair
Regulations	Good	Good	Good
Sustainability	Good	Good	Good
Intellectual Property	Fair	Good	Good

Once you have prioritized and finalized the idea, you enter the final stage—how to invent, the real development stage. This is where you determine how the development work will be conducted. Coming up with an idea isn't easy, but turning the idea into a product is even more difficult.

An idea alone is not an invention; it is only a product of your imagination, and you cannot be sure whether it will work. An invention is an idea that has been proven and put into practice. The product or process development is the stage in which you prove that idea and put it into practice.

To make any real, good progress, you need to use all your energy, knowledge, and prior experience. At this stage, you start to brainstorm again for possible answers or solutions for the final development of the product or process you have chosen. By focusing on the problem you want to solve and what the goal is, you may again come up with multiple solutions. After selecting the top solution, you will have to create a model to see if it works. Based on the results, you may end up making and testing a series of prototypes, and you'll probably come up with several designs and approaches all of which need to be screened and tested before you finalize the design.

You can create and screen your models using various methods. You can use a mental-exercise technique, drawings, or computer programs to visualize and prescreen various models. I have always preferred the mental-exercise technique since it does not require tools and can be done anytime and anywhere. This is great if you have a good memory and imagination, and if it's done right, it's the fastest way to get results. Sketching models using paper and pencil is an old method, but some people still prefer it even though you get only a two-dimensional image that makes it hard to fully visualize the actual design. The third option is to use a computer program to generate three-dimensional pictures. This comes in handy especially when you're dealing with complex designs.

Using this mental exercise, the first thing you do in the development stage is create a mental picture of what the product might look like. Depending on how good you are at creating a mental picture, you can make a lot of progress this way. I will talk about how you can develop this mental-exercise technique later on in this book. You don't learn this technique overnight—it requires long-term training and mental exercise. But once you become good at it, it becomes a very useful tool for creating and developing ideas.

I was able to complete an entire design just by using mental pictures. I started by picturing different parts for the proposed design; I imagined different ways these parts could be put together. Usually, these exercises are very fast since you can change and reimagine the pattern instantly. You can actually conjure up a three-dimensional picture of your design, and if you become good at it, you can even picture the design in action. The great thing about this technique is that you can use it anywhere and any place—even in the shower!

Once you have developed a possible design, you need to find out if it works, and that necessitates building a prototype. Most of the time, prototypes are made by hand and are very crude. You may have to go create several prototypes before you come up with a design that works.

I am particular about the cost of prototypes; I want them to be as inexpensive as possible, so I'll use hot glue, adhesive tape, rubber bands, and other inexpensive items such as plaster of Paris if you have a mold for something. A fifty-pound bag of plaster of Paris will last you a long time.

Most of what I use to construct prototypes comes from Home Depot. That's where I often find items that resemble parts of my prototype idea when I use my imagination. One time, I bought a toilet tank float since one of the parts I was looking for looked like such a float after I chopped off its top. That was one example of how my mental-picture technique worked well.

My lab had a good supply of such inexpensive materials. My friends used to make fun of me: "Charlie can't function without a glue gun and rubber bands." But I believed in conservation, and I made every effort to reuse and recycle my prototypes. Some of them became parts of other prototypes; I was always proud to make prototypes for the lowest possible cost.

My prototype for the "contraption" I mentioned that twisted a continuous sheet of paper around itself and then braided it around another twisted continuous sheet is an example of this. I wanted to make sure this prototype design worked well because I was going to demonstrate it at the annual R&D group meeting and show them my idea for creating this protective wrap.

After several visits to Home Depot, I finished the working model, which consisted of some scrap wood lying around the shop and two toilet tank floats with the tops cut off. The only expensive item in my prototype was an electric drill, and that ran all of $15. I used it to spin and braid the paper as it came out of the two rolls. The final machine was a very compact tabletop model.

Most of the top management including the CEO visited the meeting, and my prototype drew a lot of attention. I was cranking out the braided paper at a rate of two hundred feet a minute! The CEO was fascinated by that design; he kept turning the machine on and off to see if it was for real. He was shocked to find out how little I had spent—$25 in all—to make that machine, and he congratulated me for being so creative with so little.

You'll need to have a good communication with the other divisions such as sales and marketing, supply chain, finance, and so on at all stages of development. The sales and marketing team will always focus on the product features to satisfy the customer. Supply chain wants to make sure they can manufacture a product without any problems and have the capacity to supply the quantity. Finance wants to make sure the product cost stays within budget. Legal will work with you to make sure that your invention does not infringe any other patent and that you have the right to practice. Once that is cleared, it will work with you to secure the intellectual property rights. All these are long and tedious steps, and the inventor has a major responsibility in all these steps.

Example of an Invention Process

We have discussed the three different stages of inventions—identifying a problem or a unmet need, generating possible ideas that can address the problem or fulfill the need, and screening the ideas to come up with the best idea all factors considered. I will give you an example of how you go through these various stages.

Imagine you've learned that spectators at a sports game had to endure hot sun and high temperatures and some had even suffered sunstroke and had fainted. You, an inventor, see an opportunity to come up with a solution to this issue. You identify the problem—people exposed to too much heat for way too long who need some way of blocking the sun. So you generate ideas that can address the problem and fulfill their unmet need. You brainstorm and list all the ideas that result—and you remember that no idea is to be automatically deemed bad or silly.

In this case, let's say the following ideas come up and listed for further consideration.

1. Tell spectators to go home and watch future games on TV.
2. Give the people big, handheld fans to block the sun and fan themselves with.
3. Tell them to wear hats.
4. Modify the hat to create a wide shade around the head.

5. Curve the shade and make it out of dark transparent material to provide a sunglass effect.
6. Cover the top of the hat with an opaque film to completely block the sun.
7. Cover the top of the hat with a reflective film to reflect the sun away from the head.
8. Cover the top of the hat with a flexible solar panel to block the sun and at the same time generate electric power that will power a small fan that cools the face.
9. Drink plenty of water to prevent dehydration
10. Bring several cool moist washcloths in a thermal flask and use them when needed.

Once the ideas are generated, you want to analyze them for their individual value and potential development opportunity as the final solution. Let us analyze the above list one by one.

1. This is not a solution to the problem but only an alternate way to watch the game.
2. This idea has already been used as a partial solution to this problem. It will get low priority unless you can come up with a modified version possibly by combining it with some of the remaining ideas.
3. Here again, this is an existing solution, but it may be modified or redesigned.
4-7. This one along with numbers 5, 6, and 7 are simply extensions and modifications of idea number 3.
8. This one is likely to get high priority since it provides a novel approach to solving the problem.
9. This is an existing solution.
10. This is only a partial solution to keeping people cool and doesn't address the heat generation.

So number 8 moves to the top of the idea list and gets top priority for moving on to the development stage. But then you have to conduct a prior-art search before you spend a lot of time developing a product; you don't want to waste time developing something that exists. And if you find prior art, evaluate any shortfalls in it and think about modifications that would overcome them and result in a better product. Once you are satisfied with your prior-art search, you can move on to the next stage with confidence.

The last step is the actual product development. You start with the basic product concept of having some kind of hat with flexible solar panels on top and a miniature fan attached to the top capable of blowing air into the hat or onto the face. The first thing you may want to do is select a hat capable of accommodating flexible solar panels and a small fan. Once you have the components, you start your development work, and that requires imagination.

You have to come up with a design that allows the components to come together and become the final product. You can make this a mental exercise, draw things out on paper, or use a computer design program to come up with various designs. Once you have a clear picture of what the product will look like, you build a prototype. It's likely that the first prototype may not work the way you had anticipated, so you may have to construct a series of prototypes before you come up with a good working prototype. Then, you can take steps to protect your invention by initiating a patent application.

Congratulations! You've created your first invention. But you will not stop working at this stage. Inventors are never fully satisfied with a particular design; they'll imagine ways of improving it, and they think there's always room for improvement. They will be their own worst critics and focus on design features they're not completely happy with rather than on its positive features. They'll start thinking about ways to make it better, to modify it, or to expand its use for wider applications. An example in this case would be to modify the hat with a small battery for the fan to operate when it's cloudy—no sun—but hot and humid nonetheless.

An inventor may also think of other users such as construction workers or military personnel who are required to wear hard hats or helmets under extreme weather conditions. That inventor might get lucky—hard hats and helmets are much easier to handle and modify with solar panels and fans than are cloth hats.

Well, you, the inventor, just came up with an idea to expand your invention into a much larger field. Just think about the huge volume of products required in such applications!

HOW TO STIMULATE YOUR BRAIN TO PREPARE YOU TO BE AN INVENTOR

CHAPTER 6

Most of us aren't aware of our talents and what we are capable of doing. Identifying these hidden talents, nurturing them, and learning how to use them are the first steps to becoming an inventor. Just as physical exercise builds up muscles, mental exercise can stimulate, activate, and train your brain to be more creative.

Brain Exercises

Games such as crossword puzzles and Sudoku are extremely useful tools to stimulate your brain, and many computer games are designed to exercise the brain. You can actually sharpen your mind by using these tools daily. As you practice regularly, you will notice that you are getting better and faster at these games.

Another useful brain exercise is to use your mind as a blank paper and create mental pictures as if you are drawing something on paper. Close your eyes and in your mind draw a straight line. Now try to bend, twist, and turn the line to create different shapes without breaking it. Make a circle, triangle, square, or any other shape. Continue your mental exercise by imagining the letters of the alphabet you can create with that line. Some letters such as L, C, and S are simple to imagine coming out of a straight line, but mentally getting T, X, and K out of a straight line without breaking it is tougher.

Once you're comfortable with this mental exercise, start imagining three-dimensional structures. Start by placing two straight lines one on the top of the other but ninety degrees off to make a two-dimensional flat cross. Create a third line perpendicular to the cross that touches the cross at the intersection of its lines. Now, you have a three-dimensional structure. Keep adding lines in a similar way until you can visualize a three-dimensional sphere made from hundreds of straight lines or sticks. By imagining different length sticks in strategic locations, you can change the shape of the three-dimensional structure, for example, from a sphere to an egg shape. Once you have mastered this technique, you can create a mental picture of any complicated shape better and faster than you can draw it on paper. With the right degree of concentration, you can start creating any three-dimensional structure or design to get a head start on your project.

Depending on how well you progress at this mind exercise, you can slowly try more-challenging tasks such as drawing a picture, putting together a picture puzzle, or even playing chess. Many times, I have created prototype designs in my mind this way for my projects

Using Your Brain as a Database

Your brain is a powerful computer with a great deal of storage space. Most of the time, you automatically store data in your brain and retrieve it as necessary. When you meet someone for the first time, you brain automatically connects that face with a name and any other features. All the memories of that person will come back to you even the second time you hear his name. How well you recall a face depends on how much related data you have on that person when you meet. You'll remember a famous person more quickly than you will somebody who was introduced to you with only a name. In the same way, you can recall things faster if you attach more features to them.

You tend to store things you see or methods of doing things without consciously doing so. How much detail you store depends on how much attention you pay to it. For example, after using a new recipe to cook or using a new tool once, you can recall the method of using them faster than if you had simply watched somebody doing it.

You can store actions in your brain so that you can retrieve them easily and clearly when needed. Once you have stored a certain number of ideas in your brain, you can go to your database whenever you need to solve a problem and choose the right idea. In a similar way, you can also store unsolved problems with details of the challenges attached to them and retrieve them when you trigger a new idea related to a particular problem.

An invention is somewhat similar to solving a puzzle but with a major difference. For solving a puzzle, you have access to all pieces, and you need to figure out only how to put them together. But when inventing something, you don't have access to all the pieces. Inventors have to come up with the missing pieces before they can construct the whole. Many times, the invention process comes to a standstill for lack of progress in finding the missing links.

Inventors rarely work on one project at a time. Most of the time, they're involved in three to four projects simultaneously. That's because it's hard to focus on one project for a long time; they'll be forced to take a break many times due to lack of progress in the development or waiting for physical components or related information to continue the project. Rather than spinning their wheels, they move on temporarily to another project.

You cannot expect all the projects to move forward smoothly without any roadblocks; when you meet them, take a break. Keep that project in the back of your mind and let it simmer; move onto another project or a different element of your current project. Taking a break at such a time will help you to avoid tunnel vision, and chances are you'll come back to that project with a refreshed point of view and some out-of-the-box thinking. You'll need a jumpstart when you get back to that project, and you can get one by simply taking a break.

Keep all your projects in the back of your mind and periodically revisit and refresh your memory of them. Any time you see or hear something new or different, try to think about your unfinished projects and see if you can make any connections to the missing link.

Thinking Outside the Box

Inventors always challenge the status quo and look for change. They ask why, why not, and what if. They see things differently. For example, they see half of 8 not just as 4. You can slice an 8 in half and come up with two o's if you slice it horizontally, a forward-facing and a backward facing 3 if you slice it vertically, and an m and a w when you tilt those two 3s ninety degrees up and down. The point is that there may be answers other than the obvious one. I have seen many examples of seemingly obvious approaches not solving a problem and the solution coming from a nonobvious approach. But coming up with such unobvious solutions requires not getting bogged down by tunnel vision; you have to think outside the box if all you've been doing in that box is failing.

I can give you a perfect example of solving a problem by thinking outside the box. Our company used to make blood-absorbing pads placed on the meat trays displayed on supermarket shelves. Our customers complained that the absorbed blood was squeezed out of the pad when a heavy piece of meat was placed on the tray. The customer requested that we come up with a better pad that could resist the weight of the meat to prevent the blood from squeezing out. We initiated a project to improve the pad to resist the weight. I took the normal approach and tried to modify the absorbent pad for improving the absorption even under load. Some of the approaches included inserting solid objects inside the pad to support the meat and avoid direct pressure on the pad. Even though this approach worked, it complicated the manufacturing process and increased the cost of the pad significantly. I could not come up with a real solution. I continued to test other modification on the pad without much success. I thought we had run out of ideas.

We realized that we were focusing only on the pad to find a solution. The goal was to reduce the pressure applied on the pad and to prevent the blood from squeezing out. Since the customer had requested a better pad, we focused strictly on the pad. When we could not come up with an acceptable solution, we went back to look at the problem from a different angle.

It occurred to us that since we also made the meat trays for these customers, we could look at the tray design for possible solutions. We found out that it would be very easy to modify the tray to provide physical support for the meat and prevent pressure on the pad. It was very easy to mold the tray with support points without adding cost. New trays were molded with support points, and the problem was solved without making any changes to the pads. The meat rested on the support points and the pads retained the blood without spillage.

We learned two important lessons from this experience. The first obvious lesson is to have an open mind. The second lesson is that customers are not always right; we had to focus on what the customers needed, not on what they wanted. In this case, the customers wanted a better pad; instead, we gave them a better solution.

Triggering an Idea from an Unrelated Event

We continued to work on improving the meat packaging tray and the absorbent pad product. As I wrote earlier, I was working on a meat-packaging tray that would store and hide the blood coming out of the meat. Since our company made absorbent pads for that, I focused on redesigning the pads for better absorption. All the new designs showed some improvement but not to our satisfaction. So I took a break from trying to solve the problem.

As I discussed earlier, I got an unexpected clue while watching a magician pour milk into a bottle, but nothing came out when he turned it upside down. I figured out how the magician had done that and came up with redesign not of the pad but of the tray itself that would trap the liquid. The magic trick worked in real life. I came up with a working prototype in a matter of days.

Always keep unsolved problems in the back of your mind and revisit them anytime you see or hear something new or different. You'll be surprised how often you get valuable clues when you least expect them.

Nonobvious Approaches to Solving Problems

Sometimes, we rule out certain things for obvious reasons. When inventing or developing a product, do not rule out other uses for it. A good example is the use of a super-absorbent polymer (SAP) for waterproofing underwater communication cables. SAP is used in powder form in diapers; it can absorb a thousand times its weight in water. If you throw just a pinch of it into a glass of water, the entire glass of water will turns into a solid mass within seconds.

A challenge for the underwater communication cable industry was to waterproof the cable surfaces so they would last longer under water. The first step in this research project was to identify and test the most powerful water-repellent product on the market. Instead of trying the obvious materials like water-repellent products, one scientist decided to try exactly the opposite type of material—SAP—for waterproofing the underwater communication cables. Such a powerful water-absorbing material would be the last thing many people would have thought about to waterproof any surface, but his theory was that SAP that was tightly packed around the outer surface of the cable would swell and create a watertight barrier. His theory was right; he came up with the best solution to waterproofing underwater cables, and it's being used today.

Obvious Solutions Made Difficult

We look at obvious solutions when trying to solve a problem, but at times, the obvious solutions may be very difficult and may not even work. As mentioned earlier, in order to create a heat-shrink Bubble Wrap for a secure cushion product, the most obvious solution was to laminate the shrink film to Bubble Wrap. But the challenge was to laminate a heat-shrink film to Bubble Wrap without its shrinking prematurely; and the project came to a stand still.

But instead of ruling out this option, I decided to study the heat-shrink process. I found out that the film shrinks a few degrees above the softening point of the Bubble Wrap film. I had only a few degrees' window to heat laminate. This would be very tricky since this was a very narrow temperature range to operate in. But with this new knowledge, we created a heat-seal station that could accommodate this narrow sealing range. The heat-sealing process of a heat-shrink film was successfully completed without any premature shrinkage.

An inventor is trained not to rule out anything until all stones are turned over.

Taking Advantage of a Failure

You don't expect positive results every time you do research or work on a project; in fact, you probably have more failures and far fewer positive outcomes. When that happens, it is natural to be disappointed. Your success depends on how you view failure. You can give up at that point, or you can try to understand why and how you failed. Some of the most popular products we use today resulted from failed research or accidents stumbled on by clumsy scientists. Most of you are familiar with the invention of Post-it by the 3M company. Scientist Spencer Silver was trying to develop a super-strong adhesive but ended up one time with a very weak adhesive that could be lifted off easily from any surface. Instead of ignoring the failed results, he tried to find a use for this new adhesive. Later on, the famous product Post-it was developed with the help of his friend Art Fry.

Eighteen-year-old chemist William Perkin was trying to create an artificial version of quinine, a drug that treated malaria. Instead of producing the intended drug, his experiment produced a dark, oily sludge that was not even close to what he had been expecting. But he was curious about this new liquid and found out that it gave a striking shade of light purple to silk that didn't wash out. It turned out to be more vibrant and brighter than the existing dyes on the market. Those days, dyes were made mostly of insects, mollusks, or plant material. Perkins's failed test resulted in the invention of mauve, one of the first synthetic dyes.

Not all failures will be this lucky. Each failure is considered as a stepping stone toward success. It will take many failures before you find success.

Challenging the Status Quo

One reason we continue to see new and different products every day is because people are not satisfied with what they have. They always want things that look better, operate more easily, move faster, and last longer. We always hear people say, "I wish we had something better." If we did not have that attitude, there would be no reason to change things and people wouldn't have any incentive to work on new things.

Inventors have always been busy trying to develop new things to satisfy this huge appetite for new and better things. Just to see how fast things are changing, take a look at any product that we use today and compare it with an older version or even the first model to be released. Look at pictures of the first car, first cell phone, first computer, first TV, or the first of any product you're currently using—you'll be amazed at the transformation. These products have gone through many changes over the years. Today, you see everything looking better and being smaller, faster, easier to use, and cheaper.

Let us look at one example of this evolution. Many years ago, the original stool was designed as something to sit on. It probably had a round wooden seat and three legs. At that time, it was something new, and people were excited to use one; it was a change for the better. Somebody decided to make the product more sturdy by adding another leg. A back support was provided later on. Arm rests and cushions were introduced for comfort, and the stool became a chair. Later modifications were endless—wheels to move around with, recliners to relax in, ergonomic design for health reasons, vibrating features for massage, on and on.

In the late 1800s, there was talk about closing the patent office because people believed that everything that could have been invented had been invented. The strange thing about inventions is that every time you see a change, you probably think there is nothing more anybody could do to make it better, but before you know it, you see more changes. Today, technology is changing so fast that new products become obsolete before some of us even get a chance to get used to them.

When you look at new or modified products, you may think the improvement was obvious, but believe me, there was a great deal of effort and development work put into each change or modification. It is not just a matter of coming up with a new design; it is coming up with the right design that can be implemented easily and marketed at a profit.

Costs

It is amazing that we can buy new, improved products that cost less than what they replace in spite of the high cost of development, new components, and labor. The reason is that our insatiable appetite for better things demands massive volume, and mass production brings costs and thus prices down.

For inventors to be successful, they have to develop the right products the right way; they need to find out what consumers are looking for; if not, they will fail miserably. Consumers see only successful products because other products fail. According to one study, 49 percent of fast-moving consumer goods fail and fade from the market quickly.

Building on an Existing Invention

It's probable that many more inventions are created by building on existing inventions rather than coming up with something completely new. Everything including size, features, and looks can be changed for the better, and products can be used in more areas than they were originally intended for.

ENIAC, one of the first computers, took up about 1,800 square feet, required 17,468 vacuum tubes and 15,000 relays, used a teletype, weighed almost fifty tons, used 200 kilowatts of electricity, and cost about $500,000. Today, this invention has been transformed into a compact laptop capable of computing hundreds of times faster and more effectively and efficiently and at only a fraction of ENIAC's cost. It is amazing that today, pocket calculators that are more powerful than ENIAC are given away free by certain business to promote sales. This was all the result of thousands of design modification over a long time. Each one of these design modification or changes ended up in an invention.

This is the case with any of the technology or products we use today. We don't expect this technology or these products to stay the same forever; they will continually change for the better. This is true with any products we use today—cell phones, computers, cars, home appliances, and so on. There are many opportunities for an inventor to explore.

There are several ways to modify or change a product. A simple approach I used in my field was to look at the product and ask a series of questions: How was it made? Can it be made differently? What is it used for? Can we use it for something else? What if I change it by adding something to it or taking something away from it or change it so that I can add other features? Usually, you can find an answer that may lead to a new product or invention. As I have mentioned, we came up with many different uses for Bubble Wrap by asking such questions. We came up with a fire-retardant version of Bubble Wrap for the navy, we came up with anti-static Bubble Wrap for the electronics industry, and we came up with a Bubble Wrap that incorporated both of those capabilities.

People want changes for the better; inventors who understand that can develop and change things proactively. We will continue to desire new changes, and inventors are busy developing new things to feed our insatiable appetite.

Why Didn't I Think of That?

When you see something new that you like, you tend to ask, "Why didn't I think of that?" You probably could have if you thought and acted the same way the inventor did. Most of the time, people accept the idea that they are not meant to be inventors, but they could change and achieve anything they wanted if they had the desire and willingness to try. It's not easy, but people can do it with the right attitude and training. One simple way to initiate an invention is to be curious about everything you see, hear, and experience. Make it a habit to ask why, why not, and what if every time you see, hear, or experience something new.

You need to remember the basic requirements: Ambition, Imagination, Motivation, Inspiration, and Persistence (AIM-IP) described in chapter 1. All these qualities can be acquired if you don't have them already. Chapter 5 gives you a basic knowledge on how to go about inventing things. There are many things around you waiting to be invented, but first, you have to identify them. Opportunities won't come to you; you have to look for them. Your attitude is the key here.

Some people see a problem as another chore and put it aside for somebody else to work on. On the other hand, an inventor sees a problem as a challenge and jumps on it to find the answer. As the young inventor Adithyaa once said, "You don't want to be a part of the problem; you want to solve the problem."

CHAPTER 7

PROTECTING YOUR INVENTION

An invention has very little value if it is not properly protected by intellectual property law. You can come up with a great invention and think you automatically own it, but you will have no control over that invention unless you claim ownership by going through the proper procedure for protecting an invention.

Without such protection, anybody can use your invention freely, and they don't need your permission to do so. If you don't have proper documentation, it could get even worse. Others could even claim ownership of your invention by filing their own claims, and you could be left completely out of the picture. Having an unprotected invention is like leaving a bag full of cash out in the open without any protection.

The best way to protect an invention is to secure a patent. Once you have been granted a patent for your invention, you will have the full rights and ownership of that invention and nobody will be able to use or sell that invention without your permission. You can make and market your product without anybody competing with you. You can also sell your patent to others for a profit or license it to get royalty payments. You can enjoy this status for twenty years from the date of filing your patent application; after that, your invention becomes a property anybody can copy or practice. Once you have a good invention protected by a patent, you will be sitting on a gold mine.

But filing for a patent is a tedious process that requires a great deal of time, effort, and money. Most of the work in filing a patent application is done by a patent attorney, but you will need to work very closely with the attorney. You will have to make sure the attorney fully understands the nature of your invention and makes sure that all the detailed claims that you are making are included in the application. Good communication with the patent attorney is extremely important for coming up with the best patent application with broad claims for the best protection.

Prior-Art Search

Once you have an idea for an invention, you need to do some serious homework before moving forward. You have to conduct a prior-art search to see if anybody has reported the same idea or has developed something similar to your idea. You will be extremely lucky if you don't find a match. I have heard too many people saying that their idea is unique and that nobody could have even thought about it, but they tend to be wrong. Don't be surprised to see the exact idea with the same details in somebody else's work.

Not finding prior art is like winning the lottery; you are very anxious and your heart is beating fast knowing that your chances are slim but the payoff could be great. When you find a match, you feel shock and disappointment. But don't give up your hope yet; you still may get a second chance.

You need to conduct an in-depth search and go through the prior art in detail looking for any differences between your idea and the prior art. Most of the time, you can identify some differences. Many will be subtle, and you may not be able to make any significant claims for your idea. But if you are lucky, you will find some significant differences in the product itself, the process used to make the product, or some of the properties or features of the product or process. You can take advantage of these differences and take a different route by modifying your invention to get around the prior art.

Once you feel safe from prior art, you proceed with your idea to develop the product or process. During this step, you may come across different versions of your idea that lead to variations of the product or the process involved in creating it. Depending on the differences, you may have multiple inventions. In most cases, the differences will not be significant enough to be considered a separate invention.

It is important to pursue all the different versions and document them. All these variations can be included in your patent application and be protected by a patent with broader claims. You might decide to select the best version and ignore the others during the patent application, but that is the worst mistake you could make. Doing so would make it a weak patent by leaving the door wide open for the competition to take advantage of the other options; they could modify the other options to compete with your product without much difficulty. You need to include all the options and variations in your application to protect your invention with a stronger patent with broader claims. Therefore, during the development process, you need to pursue and document all the possible options before finalizing your invention.

Now you have tried to get around an existing prior art for your invention. Once your invention is on the market, the competition will try everything to get around your invention and compete with your product. Again, the best way to protect your product is to get a patent with broader claims that include every possible option you can think of. Some of the positive options that you tested may be different enough to qualify for a separate patent. Depending on the value of these options, a wise choice in this case would be to apply for a separate patent for this alternate design. Even though you are not going to exercise that option, such a patent will be very valuable in preventing others from competing with you. In industry, it is a very common practice to seek such picket-fence patents to avoid unwanted competition.

Documenting the Invention Process

You need to prepare yourself before you see a patent attorney. You need to have all the proper documentation for your work leading to the invention. The first thing to do is prepare a document for disclosing your invention. You need to provide as much information that relates to your invention as you can in this document. After going through this document, your attorney will be in a better position to make the initial evaluation and form an opinion.

You can start this disclosure document by giving a brief description of your invention including any drawings, pictures, and other details you feel are necessary. You may also include the purpose of this invention. For example, you can describe a product currently available on the market that has serious flaws and how you solved those problems with your invention.

It is also important to document the time and date when you came up with the idea. If you can, compose a brief description of the first discussion of your idea with another person as a witness and the date of such discussion. Document all the details leading to the first conception of the idea and all the steps leading to the invention. I recommend including all the tests conducted, the results of each test, and your interpretation of the results. Create charts and tables that compare your invention with similar or competitive products. You should also document and explain the differences between your product and competitive products and the unexpected results from your invention process. Documenting the dates of all the above events is also important. The document should also include all the prior-art search results and your explanation of why the prior art is different from your invention.

You need to provide your evaluation of the prior art and the justification that your invention is different from it. The attorney may or may not agree with your reasoning. Sometimes, you need to argue with attorneys to make them understand your reasoning. The attorneys conduct their own prior-art search, which will usually be even broader and more in-depth, and at times, they will come up with prior art that makes you decide to give up on your idea. You need to go through the same process of finding the newly identified differences in the prior art. Knowing all the laws, the attorney will be very strict and conservative in making a judgment before giving the green light to a patent application.

Once the prior-art search is complete and your attorney gives you clearance, he or she will start preparing the patent application with your help. Make sure your attorney has captured all your thoughts and makes the right arguments in the application. The patent application will have an abstract showing a brief description of the invention and the results of prior-art search results that show the issues or weakness with them to justify your invention as something that overcomes those weaknesses based on your data.

The most important part in any patent application is the last part in which you state all your claims. There are two types of claims made in a patent application—independent claims and dependent claims. Independent claims are the most important. These claims are stand-alone claims; the more independent claims you have, the stronger your patent will be. The dependent claims are always tied to an independent claim and have very little value by themselves.

There is a third type of claim made by a different type of patent application called an ornamental patent. The only claim for such patent will be the design that is shown by the illustration, and it protects only the particular design illustrated in that patent. That patent will not protect any variations of that illustration. Having such a narrow claim, these patents are generally considered weak.

I always had many discussions with attorneys regarding inventions and prior art. Pay close attention to the claims the attorney drafts. Experienced inventors are more likely to draft some of the languages in their claims to make sure key features of the invention are captured. Obviously, the attorney will make the final list of claims. Once the attorney fully understands the invention, he or she can expand the claims to obtain broader protection for your invention. The language used in the claims is critical and will have a huge impact on the value of the patent.

A friend of mine filed a patent application without the help of an attorney and got the patent. But that can be very risky since you could end up with a weak patent if the claims are not made properly. The claims are the most important part of a patent application. A patent attorney can help you formulate and broaden your claims; the broader the claims, the more difficult it will be for the competition to get around your patent.

When you write the claims, think like a competitor—look for any weakness in your claim and the best way to get around your patent. You can then address these weaknesses and make the claims stronger and broader. This does not mean you have to change the design at that stage; I'm talking mainly about changing the language used in the claims.

A patent with poorly written claims can be very vulnerable, and the competition can easily get around it. Simple changes in the language used in the claims can make a huge difference. For example, suppose you came up with a unique structure that has four layers of laminated film. During your development work, you initially tried two layers, and that worked fairly well, but a three-layer product worked the best. A four-layer product showed only a slight improvement in performance, and a five-layer product did not make any difference in its performance.

So at the end of your development process, you have concluded that your invention is the four-layer product and your patent claim mentions "a structure having four layers of film laminated together." You have just created a dangerous situation. If this patent is allowed, you have been granted a patent with a very narrow claim; you have protected only a structure with four layers of laminated film and thus have protection for only that. It means your competition can copy all your other structures—the two-, three-, and five-layer applications and not violate your patent rights.

Smart competitors could take advantage of the fact that the three-layered product will be less costly to produce than the four-layer product. They may even increase the thickness of the layers to match the four-layer product and its properties.

Their second option is to come up with a five-layer product that has the same properties as your product and can be made less expensive by reducing the thickness of all layers without affecting the properties. You have enabled your competition to compete with you without infringing your patent and with a minimum amount of development work. The worst scenario would be the competition getting a patent with the claims not covered by your patent.

Let's rewrite your claim—"a structure having at least two layers of films laminated together." This simple change in the wording allows you to make a broader claim and prevents the competition from making any products with two or more layers that cover the spectrum of your tested products.

I have read many patents with poorly written claims and know of competitors taking advantage of these weaknesses to their benefit or even getting patents with broader claims.

You need to be careful when trying to work around an existing invention to get patent protection for your version of a product. Let us look at a hypothetical example. You are trying to get around an existing patent with a claim for a chair having four legs. Depending on how the claims are written on the existing patent, you may be able to get a patent for a chair having three or five legs. Without looking into the details of the claims, the newly granted patent for the chair having three legs will have more value than the patent on the chair with five legs.

The reason for this argument is that the existing patent claims only a chair with four legs and therefore you are not infringing this patent if you make one with three legs. On the other hand, even though you have a patent on a chair with five legs, your patent is tied to the existing patent with the claim on four legs. Theoretically, your patent for the five-leg chair has four legs besides the additional leg. The existing patent covers the first four-leg feature in your invention. Therefore, even though you have patent on five-leg chair, you cannot practice that patent without infringing the other patent on a four-leg chair. In cases like this, it is possible that both patent owners could come to an agreement to allow them to practice each other's patents.

Before Filing a Patent Application

Before you file for a patent, keep your invention very confidential; according to US patent law, you can lose your rights to file for a patent if the invention is revealed to the public prior to filing the application; this is because your invention will be considered as being in the public domain.

If you are working for a company, you can discuss your invention with other employees on an as-needed basis. However, don't share your information with anyone outside your company if you are planning on filing a patent application. So don't start marketing a product thinking you will file for the patent later. The laws and requirements may be different in other countries.

Sometimes, the marketing team will need to understand the market potential for a product you are in the process of inventing or have just invented. Discussing an invention with potential customers may be necessary to find out its market potential and justify the cost of the product development and the patent application. You can have such communications safely without losing the rights to patent the invention, but it's best to share the information on an as-needed basis with the fewest people possible, and they should sign confidentiality agreements.

When you file for a patent, timing is equally important. Patent laws have changed. Today, patents will be granted to the first person who files for a patent on an invention, not the person who invented it first.

Filing a Patent Application for a Product vs. a Process

There are different ways to file a patent application. Suppose you came up with an idea for a product and a process to make that product. Ideally, you would file for patents on both. Depending on the nature of your invention, the patent office may ask you to file them separately. If you have to choose only one, you should definitely choose to file for a product patent because it is more valuable than a process patent for many reasons. Once you have a patent protection for a product, you can prevent others from making or selling that product; you have protection for the product regardless of how it was made. A process patent on the other hand protects only that particular process, and it is possible for others to come up with a different way to make that product.

Another difference is that it is easy to determine if others are copying your product just by looking at it, but it is very difficult to identify the process by which a product was made. You will have a tough time policing a process as it's done in privacy at the manufacturing location, so it's not uncommon for others to infringe a patent for a process. It is thus a common practice for companies not to file for a patent on a process unless there is a clear way to identify the process without actually inspecting the manufacturing location. In these cases, the process is kept a trade secret.

Inventorship

An important rule in the patent-application process is to include all those who have contributed to the invention as coinventors. Purposely excluding anybody who has contributed to the invention could lead to legal challenges. For the same reason, you cannot legally include others as coinventors if they have not contributed to the invention. If you are the sole inventor, only your name will appear on the patent. If you are the primary inventor, your name appears first followed by the name of your coinventors.

Filing a Patent Application

Once the patent application is complete, your attorney will file the application with the US Patent Office, and that can be costly. An average patent application for companies in the United States can cost as much as $25,000, and foreign patent applications can cost as much as $50,000 depending on how many countries you are filing in.

If you file as an individual, the cost can be much lower. I know people who have filed for patents for as little as $5,000, and I believe that some agencies can help you file patents for even less. The filing fee also varies depending on how many claims you are making in your patent application.

Once your patent attorney has successfully filed your patent application, it will be a long time before you hear anything from the patent office—it could be eighteen months.

After you have filed for a patent, you can market your product, but you are required to mark "Patent Pending" on the product to put others on notice that a patent application has been filed. Others can freely make and market that product during the patent-pending period, but they will have to stop doing so once the patent is granted.

Processing a Patent Application

Filing for a patent does not mean it will be granted. Patent examiners will study your application and conduct a prior-art search, and it could be much more intense; chances are, they will identify additional prior art.

Even if they don't come up with any new prior art, they will scrutinize your claims even more and under broader definitions to determine if the claims can be allowed. To be qualified for a patent application, your invention has to be novel, useful, and nonobvious. One of the rules a patent examiner would consider is, "An invention cannot be patented if the difference between the claimed invention and the prior art are such that, the claimed invention would have been obvious at the time the invention was made, to a person having ordinary skill in the art."

This definition can be interpreted different ways when applied to your claims. Usually, the examiner will have an interpretation that differs from yours and your attorney's. I have seen patent examiners rejecting claims based on this definition many times. The arguments and counterarguments between the patent examiner and your attorney can go on for a long time until one of them convinces the other and results in the rejection or acceptance of your patent application.

Usually, examiners are very tough when it comes to interpreting this definition. Chances are they will argue that your invention is obvious in light of the prior art they are citing. It is very common for the patent office to initially reject all or part of an applicant's claims, so it is very rare that you will have all your claims granted right off the bat. Only 3 percent of my patent applications were granted in the first response from the patent office. The rest of the times, the attorney had to go through lengthy arguments back and forth before finally getting the patent. In some cases, you will be required to conduct additional research before getting the patent granted.

The initial rejection can be for all or some of the claims. You should feel good if the rejection is for only some of the claims, especially if they are minor claims. Depending on the arguments made by the examiner, you will have an idea of the chances of getting those claims granted further down the road. These rejections are not final, and you will have an opportunity to make counterarguments. It is not uncommon that an examiner will make an argument that can be overcome without much effort.

Once you receive the initial rejection from the patent office, your attorney will go over all the arguments made by the examiner to reject your claims. Your involvement is very critical at this stage. Only you know all the details of your research and development work. You will be able to make certain counterarguments based on some of the minor details you thought were not important earlier. In a few instances, I was able to use such information to make a strong argument against the examiner's rejection.

Once again, documenting all the details of your research and development work is very important. After studying the arguments and the prior art cited by the examiner, your attorney will prepare a counterargument and submit it to the examiner. The examiner may accept your argument but may hit you with another rejection. The arguments and counterarguments can go on for a long time.

Many times, examiners will issue what's called a final rejection, but it will not be the end of the road; you will be able to appeal the examiner's decision to an appeal board. It will hear arguments from both sides to make the final decision. I had to go through the appeal process for two of my patents, but both times, the appeal board ruled in my favor.

Another approach is to request an interview with the patent examiner to make your argument in person. This will help you to make a better presentation of your invention to the examiner and to make sure the examiner fully understands your argument. While this is a long shot, it is sometimes worth exercising this option.

I went through this exercise once after one of my patent application received a final rejection based on prior art. I pointed out to our attorney that there was a definite difference between my invention and the prior art and that it was possible that the examiner had not fully understood certain specific features of my invention compared to the prior art.

With the attorney's recommendation, I made a physical model of my invention along with the model of the prior art. We did not expect much from this interview, but I was able to demonstrate the actual difference between the models to the examiner. I also pointed out a problem associated with the prior art and how my invention addressed that specific problem. As soon as he saw the physical differences between the two models, he agreed with our argument. The patent was granted in the next couple of weeks.

Granting a Patent

After many arguments and counterarguments, an examiner will grant the patent or issue a final rejection. If you are lucky to get your patent granted, you will be notified about the actual date of the patent grant, and you will own that patent from that day on. US patents are good for twenty years from the date they were filed; you will have sole rights to your invention during that time and can prevent others from using it without your permission.

After you have been granted a patent, you have to assert it by going after the competition that may be infringing your new patent. Even if you have printed "Patent Pending" on your product after filing the application, others can still copy your product until it is granted. You can take legal action and obtain compensation if others continue to use your invention after the patent is granted.

In most cases, all you need is for your lawyer to send an official letter to anyone infringing your patent, but it can result in a legal case that involves time and money. If you are fighting giant corporations, you are in for a long haul. They have high-powered lawyers working for them. You probably are familiar with Robert Kearns's lawsuit filed against Ford Motor Company for infringing his patent for intermittent windshield wipers. After a long and exhausting legal battle, he finally won his case and was awarded huge compensatory damages. You will find many more similar examples by doing a Google search.

Invalidating a Granted Patent

Although it rarely happens, it is possible to invalidate a patent. There is no reason to invalidate a patent unless a questionable patent issued recently threatens your business. It is not uncommon that a patent examiner could have missed some prior art or a product or process invention that was already in public domain but not widely known. In case your business is already benefiting from this prior art and the new patent somehow threatens your business, you have the right to challenge the patent based on the established prior art.

The likelihood of you winning such a case is not great since examiners are usually very thorough in their search for prior art. It is possible that the examiner was aware of the prior art but made a convincing argument against it prior to granting the patent. If the challenger can prove the validity of the prior art or the existence of the invention in the public domain, it is possible to get the patent invalidated. The only problem is that the invalidation process can be long and costly.

Other Ways to Protect Your Invention

For various reasons, people don't always protect their inventions with patents. There are other ways to protect inventions, but the protection they offer may not be as effective as a patent. There are two common reasons for not filing a patent application for an invention; the first is the danger of others copying your invention without anybody knowing it and you might have no way of policing the infringement activity. Therefore, in spite of others copying your patent, you will not be able assert your patent since you cannot prove the infringement.

The patent publicizes all the details of your invention, and everybody including your competition will have access to it. Some will try to copy it if they think they can get away with it. Most product patents are safe from being copied since it's easy to recognize any infringement by looking at the product. But patents on processes are generally vulnerable to infringement as it is very difficult to police the competition's manufacturing activity. The second reason for not getting a patent is the costs involved in getting a patent.

As mentioned before, you can also simply keep your process a trade secret, but that requires vigilance and taking all precautions to keep it a secret. The danger is that if a smart inventor comes up with the same invention independently, that person could use that invention freely without any legal issues. Depending on the circumstances, that person may even file for patent protection.

The most famous example of a trade secret is Coca-Cola's formula, which the company claims is the world's most-guarded secret. According to the company, the recipe is kept in a vault at the company's headquarters in Atlanta. If properly guarded, trade secrets are a safe way to protect your invention.

Sometimes, you can come up with an invention you are not sure about using yourself but that is valuable enough to keep from the competition, which could patent it. The best way to prevent this is to publish the information in a remote publication in a remote area where it is unlikely that the competition will notice it. If for some reason the competition comes up with the same design independently, it won't be able to get a patent since it has been published elsewhere and thus considered to be in the public domain.

CHAPTER 8
ANALYZING NOTABLE DISCOVERIES AND INVENTIONS

People often misconstrue a discovery as an invention. A discovery is recognizing something that exists but nobody has found before such as Christopher Columbus's discovery of the Americas. An invention is creating something totally new with one's own ideas and development. An invention is completely new to the world even though the physical material needed for its production already exists; the combination makes an invention unique. Thomas Edison was an inventor.

There is, however, a third category of invention—an innovation. Innovation happens when someone improves on or makes a significant contribution to something that has already been invented. Steve Jobs was an innovator.

I will take you through some notable discoveries and inventions and discuss how they came to be and what impact they have had on our lives. Learning about these inventions will definitely be an inspiration for all wannabe inventors. I think that after you learn how these inventors went about creating their inventions, you too will start thinking like an inventor.

As you read about these inventions, you will realize how they took place, what the inventors were thinking, how they worked, and how they used the information they derived from their work to create their invention. Each of these inventions is famous, useful, and valuable in its own right, so rating them would depend on personal preferences and choices. It would be tough to decide the top household inventions because the list would include telephones, televisions, refrigerators, microwave ovens, washing machines and dryers, vacuum cleaners, and so on. All these inventions have become part of our lives, and we would probably find it very hard to not have them. And depending on how often we use these, we would value and thus rate them differently. But they are all valuable, essential, and great inventions. From the point of view of an inventor, you may see more value in one than the others depending on how it led to other inventions.

Ancient Discoveries and Inventions

I was curious to know what humanity's first discovery was. My Google search indicated that the earliest knives used by humans were probably sharp rock flakes or water-worn creek cobbles. The actual invention is presumably handmade wooden scrapers or wooden chopping devices that were first used more than 2.6 million years ago. Eventually, these wooden knives gave way to stone and then metal knives.

Fire may be considered an old invention though it was discovered rather than invented. The actual invention was how humans figured out how to create, use, and control fire.

Prehistoric clothing is considered an ancient invention. The earliest clothing was probably made from leaves put together and wrapped or tied around the body. Eventually, more-convenient animal leather or fur was used for clothing. Archeologists have identified very early sewing needles of bone and ivory from about 30,000 BC that were probably used to stitch leather. We still use fur coats as high-end clothing—it seems like nothing has changed.

Naturally occurring pigments are considered another ancient discovery dating back 400,000 years. Later on, pigments such as ochres and iron oxides were invented and used as colorants. There is evidence that early humans used paint for aesthetic purposes such as body decoration. Later on, pigments from unusual sources such as botanical materials, animal waste, insects, and mollusks were created.

Archaeologists have identified the use of log boats as early as

60,000 years ago. Evidence shows that the early Egyptians knew how to assemble planks of wood into a watertight hull.

The first known music instruments were flutes. Archaeological data indicate that a prehistoric bone flute was identified as early as 30,000 years ago. A three-hole flute made from a mammoth tusk was excavated from the Geißenklösterle cave in Germany. Some early flutes made from the wing bones of red-crowned cranes with five to eight holes each were excavated from a tomb in Jiahu in the Central Chinese province of Henan.

Ancient Egyptians are known for making the first rope for fastening, pulling, carrying, and hunting. The earliest ropes were probably made from plant fibers such as vines. Eventually they figured out how to make the actual rope by twisting and braiding these fibers together. Egyptian rope dates back to between 4000 and 3500 BC and was generally made of water reed fibers. Eventually, ropes were made from leather and animal hair, and this technology led to the modern yarn used for clothing.

History shows that the first wheel probably originated in ancient Sumer (modern Iraq) around 5000 BC. The wheel reached India and Pakistan in the third millennium BC.

Let us look at some of the selected early inventions in chronological order.

Compass

The earliest compasses were made of lodestone in China between 300 and 200 BC. This first navigational device has been a major force in human exploration. We came a long way to create the modern GPS system.

Paper

Paper was invented around 100 BC in China, and it has become an integral part of human life as it is used for writing, printing, packaging, and cleaning. It seems that paper usage has peaked. A major emphasis on recycling and reusing has made an impact on conserving and reducing the use of paper.

Many more inventions came out of the effective reuse and recycling processes of paper and pulp. Today, the world is witnessing the beginning of the paperless era. Thanks to the computer age, industry is slowly turning into a paperless environment. As a result, we are cutting down fewer trees for making paper pulp and thus using less energy for paper manufacturing.

Gunpowder

The invention of gunpowder has changed the world. Chinese scientists invented the gunpowder by accident while experimenting with life-lengthening elixirs around AD 850. They had been experimenting with a powerful oxidizing agent, potassium nitrate (saltpeter), in medical compounds for centuries. In one test, they mixed it with sulfur and charcoal, which resulted in an explosion that burned down a house. Their explosive invention became the basis for almost every weapon from fiery arrows to rifles, cannons, and grenades.

Optical Lenses

Optical lenses were another early invention that greatly influenced our lives. Egyptians first developed them with contributions from Mesopotamians and ancient Greeks later on. This invention led to further developments of glasses and microscopes, telescopes, and other optical lenses. Optical lenses were instrumental components in the creation of media technologies including photography, film, and television.

Printing Press

Printing presses laid the foundation of our modern age in many ways. Johannes Gutenberg, a German scientist, invented the first printing press in 1439; he introduced mechanical, movable-type printing. The speed with which pages could be printed revolutionized the way knowledge was spread compared to the slow process of writing and copying books by hand.

Electricity

The discovery of electricity was the result of a slow evolution by the contribution of a number of bright minds over many years. The origin of this work goes back to ancient Egypt and ancient Greece, although Benjamin Franklin is generally credited with significant contributions to our understanding of electricity. This discovery has profoundly changed the way we live, and today, we couldn't fathom the idea of our world functioning without it. A major extension of this discovery was the invention of the lightbulb by Thomas Edison in 1879.

Steam Engine

James Watt invented the first steam engine around 1770. For decades before his invention, many attempts were made by other scientists to build steam engines but without much success. James Watt ended up using some of the ideas of his predecessors to successfully complete his steam engine. Its impact was enormous, as it led to steam-powered ships and trains that could transport large volumes of goods faster and cheaper than existing methods.

Telephone

Scottish-born American scientist Alexander Graham Bell was granted the first patent for an electric telephone in 1876. There were controversies about whether he was actually its sole inventor, and he faced years of legal challenges to his claim that resulted in one of the longest patent battles.

He worked at a school for the deaf, and he wanted to invent a machine that would transmit sound by electricity. According to Bell, "If I could make a current of electricity vary in intensity precisely as the air varies in density during the production of sound, I should be able to transmit speech telegraphically." The first coherent complete sentence, the famous "Mr. Watson, come here; I want you," was transmitted in his laboratory. Although Bell was the first to design and patent a practical device for transmitting the human voice by means of an electrical current, he always described himself simply as a teacher of the deaf.

Vaccination

Edward Jenner and Louis Pasteur were medical pioneers. Jenner, a country doctor in England, noticed that milkmaids were often infected with cowpox visible as pustules on their hands or forearms. He also noticed that they were immune to subsequent outbreaks of smallpox that periodically swept through the area. By applying the scientific methods of observation and experimentation, he conducted one of the world's first clinical trials. He succeeded in a controlled transfer of pus from one person's active smallpox lesion to another person's arm. Jenner developed the first vaccine for smallpox in 1796.

Louis Pasteur, a French chemist and biologist, developed the rabies vaccine in 1885. Today, vaccination is a major part of medicine and it has eradicated many deadly diseases.

Car

Although many individuals were responsible for inventing cars, German Karl Benz was credited for creating what is considered the first practical motorcar in 1885. The modern car with all the new features and luxuries is the result of many thousands of inventions and patents. It is interesting to note the evolution of the modern car from the very first design—from hand-cranked ignitions to the remote-start engines and from power steering to the latest self-driving cars.

Airplane

The Wright brothers' invention of the airplane was the result of exhaustive research and many design trials. The history shows that they tested over 200 wings and airframes of different shapes before finalizing the design. On December 17, 1903, Wilbur and Orville Wright made four brief flights with their first powered aircraft. They piloted the gasoline-powered, propeller-driven biplane, which stayed aloft for twelve seconds and covered 120 feet during its inaugural flight. That was the beginning of the modern aviation age.

The Wright brothers continued to develop new designs, and in 1905, their aircraft could perform complex maneuvers and remain aloft for up to thirty-nine minutes at a time. In 1908, they traveled to France and made their first public flight.

Their prototype became the foundation for many more advanced designs. Today, we are breaking all the flight records by flying around the world nonstop and faster than the speed of sound.

Shivkar Bāpuji Talpade was an Indian scholar who is said to have constructed and flown an unmanned airplane in 1895, ten years before the Wright brothers' invention.

Penicillin

One of the most significant breakthroughs in medicine was the discovery of penicillin's ability to cure infectious bacterial diseases. Today, antibiotics are an integral part of modern treatment. They are compounds produced by bacteria and fungi capable of killing or inhibiting competing microbial species. Apparently, this phenomenon has long been known—the ancient Egyptians applied poultices of moldy bread to infected wounds.

Alexander Fleming, a professor of bacteriology at St. Mary's Hospital in London, revolutionized all medicine when he accidentally discovered penicillin, the first true antibiotic, in 1928. Fleming began to sort through petri dishes containing colonies of *Staphylococci*, a bacteria that causes boils, sore throats, and abscesses. He noticed something strange in one dish. It was dotted with colonies except in one area where a blob of mold was growing. The area immediately around the mold was clear, and it looked as if the mold had secreted something that inhibited bacterial growth. It was later identified as a rare strain of *Penicilliumnotatum*.

https://www.acs.org/content/acs/en/education/whatischemistry/landmarks/flemingpenicillin.html.

Anesthesia

The name of the original discoverer of anesthesia has been debated. Crawford Long, William Morton, Charles Jackson, and Horace Wells were credited with realizing the effect of ether and nitrous oxide in inhibiting pain. In those days, inhaling either of these compounds was somewhat popular for recreation and entertainment. Wells attended a party at which someone had injured his leg while under the influence of laughing gas. The man, whose leg was bleeding, told his friend that he didn't feel any pain. Based on this accidental discovery, Wells decided to try this on himself by removing his tooth after using this compound as an anesthetic. That was the start of using anesthesia during medical procedures and surgeries.

Nuclear Fission

Nuclear fission is the process by which a heavier atom splits into lighter atoms and releases considerable energy, and it has had a profound effect on our world. The pioneering work of Otto Hahn, Lise Meitner, and Fritz Strassman was a crucial step in the long scientific journey that led to the development of nuclear technology. Scientists quickly recognized that if the fission reaction also emitted enough secondary neutrons, a chain reaction could potentially occur and release enormous amounts of energy. Many scientists joined the efforts to produce nuclear reactors and atomic bombs.

Semiconductor

A semiconductor is a substance, usually a solid chemical element or compound, that can conduct electricity under some conditions but not others, and that makes it a good medium for the control of electrical current. The discovery of semiconductor materials allowed tremendous and important advancements in electronics. We needed semiconductors for the miniaturization of computers and computer parts and for manufacturing electronic components such as diodes, transistors, and many photovoltaic cells.

Anything that is computerized or uses radio waves depends on semiconductors. Semiconductors are not an invention, and no one invented the semiconductor, but many inventions use semiconductor devices. Michael Faraday was the first person to observe a semiconductor effect in 1833. Faraday observed that the electrical resistance of silver sulfide decreased with temperature. In 1901, the first semiconductor device patented was called "cat whiskers." Jagadis Chandra Bose invented this device. A cat whisker was a point-contact semiconductor rectifier used for detecting radio waves.

Computer

It is hard to determine who invented the computer because of the many different classifications of computers. The word *computer* was first recorded as being used in 1613 and originally was used to describe a human who performed calculations or computations.

That definition of a computer remained until the end of the

nineteenth century, when the Industrial Revolution gave rise to machines whose primary purpose was calculating. Charles Babbage created the first so-called mechanical computer in 1822, but it did not resemble modern computers. In 1822, Babbage conceptualized and began developing the Difference Engine, considered the first automatic computing machine.

The Z1 was created by German Konrad Zuse in his parents' living room between 1936 and 1938. It is considered the first electromechanical binary programmable computer and the first really functional modern computer.

J. Presper Eckert and John Mauchly at the University of Pennsylvania invented the first electronic computer used for general purposes known as ENIAC (Electronic Numerical Integrator and Computer). ENIAC occupied about 1,800 square feet, used 17,468 vacuum tubes and 15,000 relays, used a teletype, weighed almost 50 tons, used 200 kilowatts of electricity, and cost about $500,000.

Altair 8800, the first "personal" computer, was introduced by Ed Roberts in 1975. (Some consider KENBAK-1, introduced in 1971, as the first personal computer, though it was not called a personal computer). IBM introduced the first portable computer in 1975; it weighed fifty-five pounds and had a five-inch CRT display. It was tape driven and had a 1.9 MHz PALM processor and 64 KB of RAM.

World Wide Web

The worldwide network of computers has been in development since the 1960s. Tim Berners-Lee is credited with creating the web in the 1990s. Today, we depend on the internet for practically everything, and it is the first thing we turn to when we need to communicate or when we need information on anything be it commerce, entertainment, or politics.

CHARLES KANNANKERIL

Early African-American Inventors Who Changed the World

Thomas L. Jennings (1791–1859) was the first African-American to receive a patent in the United States. He paved the way for future African-American inventors to gain exclusive rights to their inventions. He lived and worked in New York City. He invented an early method of dry cleaning called "dry scouring" and was granted a patent for it in 1821.

At that time, the US patent law stated, "[a slave master] is the owner of the fruits of the labor of the slave both manual and intellectual." Jennings, however, was a free man and thus received his patent. It was reported that Jennings used the money from his invention to free the rest of his family and to donate to abolitionist causes.

In 1912, **Garrett A. Morgan** invented the safety hood, which later became the gas mask that we use today. His design involved a hood placed over the head and a long tube extending from it to reach beyond the gas fumes. A moist, spongy material was stuffed into the end of the tube to prevent smoke and dust from entering the tube. The hood was also attached to a hose to breathe out of. This hose was connected to a one-way valve to prevent smoke from entering the hood so the wearer could breathe. After several successful demonstrations on how this device could save lives, the news about this new product spread across the country, and police and fire departments started using them.

Morgan started his own company, and the orders poured in. It was reported that in the Deep South, orders for this product stopped abruptly when they found out that the inventor was an African-American. Some of the other famous inventions by Morgan were the traffic light and hair-straightening cream.

George Washington Carver (1860s–1943) was born into slavery in Missouri. After the Civil War, he received an education. He became famous for conducting research in agriculture and teaching farmers about fertilization and crop rotation. He was also credited with discovering many uses for peanuts.

THE INVENTOR IN YOU

Jan Ernst Matzeliger came to this country as an immigrant from Dutch Guiana. He worked as an apprentice in a Massachusetts shoe factory. In his day, shoes were made mainly by hand, and not too many people could afford them.

While working at the shoe factory, Matzeliger invented an automated shoemaking machine that attached a shoe's upper to its sole. His invention resulted in the mass production of shoes that the average person could afford.

In 1867, **Alexander Miles** invented and patented the first automatic doors for elevators. This invention prevented many unfortunate accidents. Today's elevators still employ similar technology.

Elijah McCoy invented the lubricating cup for train engine gears. Before his time, trains had to stop frequently so engineers could lubricate the gears by hand, and McCoy's inventions automatically handled the lubrication. This significantly improved efficiency in the locomotive operation. It is worth mentioning that Elijah McCoy is the man behind the phrase "The Real McCoy."

Andrew Jackson Beard (1849–1921) was born into slavery Alabama and gained freedom when he was fifteen. During the early part of his life, he invented a flour mill, a rotary steam engine, and two kinds of plows. He decided to work for the railroad in the 1890s. In those days, workers had to insert a metal pin to link rail cars; that was a very risky operation and caused many accidents. While working for various railroad companies, Beard created his most famous invention, the Jenny coupler. His new invention automatically locked train cars together when they bumped into each other. This made connecting long trains for travel and trade much easier and safer.

Lewis Latimer (1848–1928) joined Thomas Edison's research team and became the head draftsman for General Electric. In 1882, Latimer invented a carbon filament to use in lightbulbs. It lasted longer and was cheaper than Edison's first design. Latimer's inventions include a bathroom for railroad cars, a disinfecting and cooling device, a hat and coat rack, locking umbrellas, and a device for supporting books.

George Washington Murray (1853–1926) had a real interest in farming. His work resulted in eight patents related to farming. Murray, born a slave, was elected to the US House of Representatives from South Carolina in 1892.

Accidental Discoveries and Inventions

Many inventions were the result of accidents. Some of these accidents were caused by absentminded or clumsy scientists. The important thing here was how they viewed these accidents and how they turned the results into something useful instead of ignoring these accidents.

X-ray

German physicist Wilhelm Röntgen discovered the invisible power of X-rays by accident. He was experimenting with cathode-ray tubes (glass tubes with the air sucked out and a special gas pumped in). When he ran electricity through the gas, the tube glowed. When he turned on the machine after covering it with a black cardboard, he noticed that a chemical placed a few feet away started to glow. In spite of blocking the visible light, something else had passed through the black board. The cathode-ray tube had been sending invisible rays that could pass right through paper, wood, and even skin. The lab chemical that lit up reacted to these rays. He called the phenomenon X-rays—the X stood for "unknown." The contribution this invention made to the medical industry is immeasurable.

Analyzing the accidents or strange phenomenon during their research work is a pattern you see in most inventors. In many cases, at that point, their attention is diverted from their real goal, and it often ends up in the creation of even greater inventions.

Plastic

Leo Baekeland, a Belgian-born chemist, was trying to find a replacement for shellac, a resin secreted by a South Asian scale bug. He tried a combination of formaldehyde and phenol (an acid extracted from coal tar), but it failed to catch on as a substitute for shellac. After several failed attempts, he tried to mix it with various fillers such as wood flour and slate dust by controlling the temperature and pressure.

In 1907, he created a moldable material that was tough, nonconductive, and heat resistant. The container he used for this test was a massive iron cooker that he called a bakelizer, and therefore he called his invention Bakelite. His invention was widely used to make electronic components, auto parts, telephone casings, and many thousands of items for decades.

Bakelite opened the door to other synthetic plastic materials such as Plexiglas, polyester, vinyl, nylon, polyurethane, polycarbonate, and so on. Today, pretty much everything you touch is made from plastic—from auto parts to ziplock bags, A to Z.

Unfortunately, our world is saturated with plastic products that can remain in the environment for centuries without degrading. We all need to do our part to address this serious pollution issue by reusing plastic and supporting recycling efforts. This is a real need that has been identified, and there are many opportunities for invention that would address it. It is ironic that we are now looking for an invention to solve a problem created by one of the greatest inventions of the past. As an inventor, you can take this challenge and see how you can contribute to save our environment.

Saccharin

In the 1870s, a Russian chemist named Constantin Fahlberg was experimenting at Johns Hopkins University with coal-tar derivatives to determine how they reacted to phosphorus, chloride, ammonia, and other chemicals. One night, he was eating dinner after work when he realized his dinner roll tasted unusually sweet. He soon realized that his hands were covered with a chemical that made everything sweet. He had happened to spill an experimental compound over his hand during work that day. He ran back to the lab and tasted everything on his worktable. The sweet taste came from an overboiled beaker containing what became the zero-calorie artificial sweetener called saccharin. Fahlberg probably would have never discovered saccharin if he had washed his hands after each lab session as normal chemists do.

This artificial sweetener changed the food industry, and it allowed millions to enjoy sweet food without consuming sugar. His accidental observation, curiosity, and inventive mind led to the development of a breakthrough product in the food industry.

Teflon

Twenty-seven-year-old Roy Plunkett invented Teflon while working on developing a better refrigerator. He wanted to combine hydrochloric acid with tetrafluoroethylene gas. Before mixing, he cooled and pressurized the gas in canisters overnight. The next day, he noticed that the gas had disappeared from the canister. Nothing came out when he opened it although the canister weighed the same as it had when it had been full of gas. Curious, Plunkett cut the canister in half and noticed that the gas had solidified and formed a slick coating on the inner surface of the canister. He pointed this out to the team, and rather than ignoring the apparent failure, they decided to test the new material. They soon realized this new polymer had very unusual properties—it was extremely slippery and was resistant to most chemicals including corrosive acids. In 1945, DuPont trademarked this product as Teflon. Although Plunkett invented Teflon, the idea of using it for the cooking pans came from French engineer Marc Grégoire years later.

Inventors always pay great attention to the results of their work whether successful or not. They don't normally ignore failures; instead, they get curious and want to know why and how it happened. This often leads to many inventions.

Velcro

Swiss electrical engineer George De Mestral was walking in the woods with his dog one day and came back with cockleburs clinging to his clothing and his dog's fur. He was curious to find out why it was so difficult to remove them from his clothes and his dog's fur. He looked under a microscope and noticed hundreds of tiny hooks that line the surface of the cocklebur and discovered they could easily attach to the small loops found in clothing and fur. He decided to develop a mechanism similar to the cockleburs' hooks and loops.

In 1959, after experimenting with different materials, De Mestral developed Velcro using nylon polymer. The name Velcro came from the combination of the words *velvet* and *crochet*. In the 1960s, NASA made it popular when Apollo astronauts used it to secure items they didn't want escaping in their zero-gravity environment. Today, Velcro is widely used as a fastener by many industries including hospitals and athletic companies including Adidas, Reebok, and Nike. https://www.quora.com/what-are-some-accidental-invention-stories-that-most-people-dont-know.

Besides observing and identifying things, inventors are very creative in using that observation to come up with practical solutions.

Popsicle

Eleven-year-old Frank Epperson invented the Popsicle purely by accident in 1905. He often made a popular concoction, a fruit-flavored soda, out of powder and water. One evening, he left a batch of unfinished soda outside overnight. The stirring stick was still in the cup, and the night got very cold. The next morning, he discovered that the drink had frozen around the stick. He called the delicious treat the Eppsicle.

Later on, he made Eppsicles for his children, who called it Pop's Sicle. Years later, he served the frozen lollipops to the public at a fireman's ball, and they were a huge hit. In the following years, he enjoyed even more success after serving them at Neptune Beach, an amusement park in Alameda, California.

Epperson applied for a patent and began producing more fruit flavors. He sold the frozen pops on birch sticks for a nickel apiece. Today, people all over the world love Popsicles.

Many inventors are also innovators. They will have the vision for the future. They improve their inventions and continue to make contributions.

Matches

While stirring a pot of chemicals, John Walker noticed that a dried lump had formed on the end of the mixing stick. When he tried to scrape of the dried lump, it ignited suddenly. Based on this observation, he developed the first matchstick. He started to make matchsticks and sold them in a box with a piece of sandpaper for a striker. Walker wasn't interested in patenting the idea, so others copied his idea and manufactured and marketed even better products.

Invented by Johan Edvard Lundstrom, modern safety matches are made with nonpoisonous red phosphorus. The Diamond Match Company was the first to sell such safety matches in the United States. They forfeited their patent rights to allow all match companies to produce safe matches.

Invention and innovation are not the same. Not all inventors are innovators. John Walker, who invented the first matchstick, did not want to think beyond his invention and made the mistake of not protecting his invention. That meant others were free to copy and modify it to create a better and more successful product.

Viagra

Simon Campbell and David Roberts, two researchers at Pfizer, were working on developing a new drug they hoped would treat high blood pressure and a heart condition called angina. The development work was completed in 1980. The results of the clinical trials were disappointing as it was not as effective as Campbell and Roberts had predicted.

While analyzing the failure, they looked at the side effects of the trial. Many patients underwent this trial reported that the treatment lead to erections. Their research team decided to focus on this side effect, and the company launched a new clinical trial to use the drug for erectile dysfunction. The trial was a huge success, and the drug, named Viagra, was approved by the US Food and Drug Administration in 1998.

The curious minds of inventors will never let them ignore a failure. A failure can be a stepping stone either toward your project's success or the start of a new adventure.

Post-it Notes

Spencer Silver, a chemist for 3M, was working on a project to develop a strong adhesive for the aerospace industry. One of his tests produced an entirely opposite product—a weak adhesive made of tiny acrylic microspheres. The spheres were very durable and would stick rather well even after several uses. Nobody paid attention to his product because it seemed useless. He suggested selling the adhesive as a sticky surface for bulletin boards to post and remove notices without nails or tacks, but he didn't get any support from the company.

Years later, Art Fry, another 3M chemist, identified a use for this adhesive. He was a frequent choir singer, and he didn't want the bookmarks he used in his hymnal to slip out each time he opened it. He thought about the new adhesive his coworker Spencer Silver had come up with. His idea was to put the adhesive on the paper instead of applying it to a bulletin board as his friend had suggested; that way, he could stick the paper with this adhesive on it to anything.

They approached 3M management with this idea, but neither management nor the marketing department backed it. It appears that the marketing department in this case may not have conducted a proper market evaluation with this invention and did not see the opportunity.

Years later, one of the lab managers decided to market it themselves without the marketing department's support. The rest is history.

There are some important lessons to be learned here. You don't necessarily fail if you don't get the results you're looking for. It is not the results but what you do with the results that is important. Spencer Silver got the opposite results he was looking for, but instead of ignoring it, the team turned a failure into a fortune.

Silver thought only about putting the adhesive on a bulletin board while Fry broadened his vision and came up with the idea of putting it on paper. Avoiding tunnel vision and thinking outside the box are important factors for a successful invention.

Silly Inventions That Made Millions

Inventions do not have to be very scientific, technical, or sophisticated to be successful. History showed that many simple and silly ideas became very popular and the inventors made millions of dollars from these ideas. Consider these examples.

Hula Hoop

This was an incredibly simple toy. The designers of this toy got the idea while watching children playing with bamboo hoops. The designers decided to mass-produce the toy using plastic.

Wacky Wall Walker

Ken Hakuta received a Wacky Wall Walker toy as gift from his mother. He was fascinated by the way the toy would crawl down when thrown against a wall. He believed in this product and decided to buy the rights from the original Chinese inventor for just $100,000. He started to market this toy in the United States, and it became a huge success. He made more than $80 million in sales.

The Top-Down Squeeze Bottle

These are commonly used in restaurants for dispensing ketchup. The inventor, Paul Brown, came up with his signature silicon valve in 1991 that would open when the bottle was squeezed but would close once the pressure was let up. He sold his invention to Heinz, shampoo manufacturers, and even NASA. He sold his company along with the rights to the design for $14 million.

Pet Rock

Gary Dahl created this simple idea for a toy—simply a stone placed in some hay and packaged in a fancy container. His marketing success was to offer a hassle-free pet and to target people who didn't have enough time or energy to look after a real pet. His manufacturing cost was negligible, and he was able to sell this product for $3.95—making almost $15 million in just six months.

Slap Bracelets

Stuart Andrews invented these, and all the kids loved them. Most of them had several and would slap them on all at once. Andrews hit the jackpot with this idea. He was a high school shop teacher before he invented the slap bracelet. In 1990 alone, the bracelets were estimated to be profiting between $6 million and $8 million.

Slinky

Naval engineer Richard James invented this as the result of a clumsy accident. After dropping a tension spring he was working with, he watched it slink away across the floor. He decided to use this idea for a toy. Later, he manufactured and marketed this toy and sold it for $1. It enjoyed one of the fastest-ever growths in sales with an estimated profit of $250 million.

Inventions of the Future

We have just seen how certain things were invented and how they changed the way we live. Try to imagine what this world will look like tomorrow. Things changed rather slowly in the past, but today, the pace of change is rapid; it's in high gear. We don't have to wait long to look back and see how things have changed. It was a long time from the telegraph to email, but it was a very short time from email to texting.

Driverless cars are here, and they will probably evolve into self-driven taxis, trucks, and buses. Stores, which have been around forever, are giving way to online shopping. Robots are being designed that can do anything we can do. The way things are changing, we may not have any reason to leave the house. These inventions will dramatically change our lives. New inventions and disruptive technologies could change the world into something we may not recognize. Our future lifestyle may seem very unsettling.

3-D Printing

We have already been exposed to 3-D printing—the process of creating anything by printing—and that includes clothes, food, toys, or parts for anything we use. A document containing the structural details of the object to be printed is fed into the machine that prints out the 3-D structure. We have only scratched the surface of this new technology, and it will have a huge impact on the way we make things.

Robotics

Robotics is another exciting innovation. This technology has advanced to a level previously thought to be too difficult or expensive to automate. We have already seen robotic surgery and prosthetics. The use of robots in industry is very common. The future may be in nanorobotics, robots of an incredibly small size.

Medicine

The medical field is another area in which we will see many innovations. We could expect to see new vaccines, genomic-directed clinical trials, and new cures for deadly diseases such as cancer. Remote physical examination and health monitoring could revolutionize the medical industry.

Artificial Intelligence

The concept of intelligence created by computers rather than people is an area that is under increased focus. This type of artificial intelligence is used when machines copy the cognitive functions of the human brain in learning and solving problems. Basically, this means we could see computers and machines that could think and react like human beings.

Space Exploration

Space exploration and colonization is another area in which we can expect many changes. Soon, trips to space will be very common, and settling in planets and moons may not be far-fetched.

CHAPTER 9

ROLL UP YOUR SLEEVES—IT'S YOUR TURN TO INVENT

I hope by now that you have learned the criteria for an invention and how to go about creating one. It is time for you to think like an inventor. I encourage you to take the challenge and see if you can come up with an invention with the guidance provided in this chapter.

To recap, in the first chapter, you learned about the criteria for an invention and the basic requirements for an inventor. The second chapter described how thinking differently will make a difference. The third chapter talked about the examples of thought processes that led to many inventions. Chapters 4 through 8 covered how ideas are generated, different steps in the invention process, how to stimulate your mind and motivate yourself to be an inventor, and how to protect your invention.

We also looked at some notable inventions starting from ancient discoveries to modern technological inventions. These examples are inspiring to anybody with the desire to invent something. As you can see from these examples, inventors come from all walks of life, and you don't have to be a genius or a scientist to be an inventor; anybody can become one. No form of handicap is a reason not to become an inventor. Louis Braille, a sixteen-year-old-boy, invented Braille, and he was completely blind.

Before you start, let us go over one more time the basic principles of the invention process. Remember the formula AIM-IP—AIM for Intellectual Property—and make sure you have a comfortable level of Ambition, Imagination, Motivation, Inspiration, and Persistence.

As you prepare yourself, change your attitude a bit and start challenging the status quo. If you are satisfied with the way things are and the things you use, you probably won't find a reason to invent anything. Try to be critical of everything; instead of seeing only the good in everything, look at the negative and identify the areas that could be improved. This is one way you can lay the foundation for your first or your next invention, so develop a critical mind-set.

Once you do, you can get the ball rolling with your curiosity about the things around you, especially new things. Start asking questions—Why? Why not? What if I do this?—every time you see or hear something. Remember, always try to have an open mind—think outside the box. These simple steps will help you progress through your invention process.

While writing this chapter, I gave some homework to Mary, my wife, to Luke, my seven-year-old grandson, and to Lena, my four-year-old granddaughter. I asked them to think about the things they do or use every day. Their assignment was to pick an item they would like to see different, better, or improved and come up with something specific they wanted to see changed and how. I gave them a couple of days to think about it. The following are their responses.

My wife immediately responded by identifying a problem; unfortunately, I happened to be responsible for it. She often finds the toilet seat up when she wants to use it. She has to manually bring the seat down with a tissue to avoid touching it. To solve this problem, she wanted a remote-controlled device that would manage that.

Mary identified a common problem we see in most households. Inventors can approach this problem in many ways. They could resort to a simple mechanical design—a foot pedal that controls the toilet seat to move up or down. Or they could apply high tech—an app in a cell phone that could control the seat position.

Luke told me he would like an improved backpack that would allow him to retrieve something from it even when he was walking and it was on his back. He wanted a backpack that could be flipped forward even while he was en route to school. He identified a real problem for schoolchildren or even adults who carry backpacks. The possible solution he suggested had a lot merits. He provided a good start for an inventor to design a backpack that solves this problem.

Lena told me she would like to have more fun riding her bike by being able to listen to music from a cell phone attached to the handlebars, maybe with a rubber band. Lena identified an unmet need and a possible solution. An inventor who looks at this scenario could come up with a new product to attach the cell phone or any music-providing gadget to the handlebars.

Having learned all about principles of invention and its process, I believe going through a similar exercise would be a great start. I encourage you to get involved and participate in the following homework. By doing so, you can actually identify several unmet needs or problems you can work on that could lead to an invention. Besides identifying these new areas to invent, you can also come up with a few ideas for possible solutions and develop the confidence to embark on a new project. Good luck!

Identifying Opportunities to Invent

Your first assignment is to generate a wish list after you engage in a brainstorming session. I will walk you through some areas in daily life and give you hints about how you can come up with your own list. As you go through these items, list things you don't like about them and want to see changed. You can even prioritize these features under "must have," "like to have," and "nice to have." Feel free to add more items to the list below.

I have also listed some of the common comments we hear about these items; these are examples that can go on the wish list, but go over this entire list, make comments, and generate your own wish list. You don't need to sit in front of this book and think all the time. Keep the list in the back of your mind and recall it every time you see, hear, or do anything. You don't have to stick with this list, and ideas may not come to you right away, so be patient. Keep thinking about the challenge, and ideas will eventually come to you. In this brainstorming session, you can bring up anything you want. Remember—don't judge any ideas as bad or silly. You will be surprised how many items will end up on your wish list. This will give you plenty of reasons and opportunities for you to start thinking about your own next invention.

Home Appliances

Vacuum cleaner: Common complaints are that vacuum cleaners are too heavy, too noisy, are not designed ergonomically, have suction nozzle adapters that are not shaped to clean corners and difficult-to-reach areas, or that it is difficult to change and clean the filters or the bag. Customers would like one nozzle for all types of cleaner. It would be nice to have a vacuum cleaner capable of multiple functions—vacuuming, wet mopping, waxing, etc.

_____

_____

_____

Washing machine and dryer: Consider a washing machine with a built-in and programmable detergent, bleach, and fabric softener dispenser. Is there a single machine that does both the washing and drying in a single load?

_____

_____

_____

Refrigerator: Is it possible to design a fridge capable of identifying and giving warning signals about expired food products? Indicators on the fridge could let you know when you are running out of certain food items. How about programming a slow conveyance system that moves frozen food from the freezer section to the refrigerator section thawed and ready to cook when needed? Water dispenser on the fridge could provide both cold and hot water.

_____

Microwave oven: Consider a programmable cooking cycle with recipes, a single unit for a conventional and microwave oven, or a microwave oven capable of handling metal containers.

Dishwasher: Can a dishwasher be invented that is capable of cleaning dishes using steam in place of detergents?

Portable fan: Consider a portable fan capable of producing extra cool air (with mist?) for summer and able to blow hot air during winter.

Household Items

Shaving cream dispenser: Shaving cream is usually dispensed in the hand and then applied to the face. Would it make it easier to attach a brush (like a paint roller brush) to the shaving cream can to dispense and apply the cream in one step?

Shower: Consider the safe use of hot air (similar to hand dryers in public washrooms) to dry the entire body after the shower is turned off instead of using towels. It could help save the environment by reducing the number of towels in the laundry. What about a shower with support poles and remote-control buttons for the handicapped?

Dental hygiene tools: Think of new designs for better dental hygiene—toothbrush with on-demand toothpaste dispenser, toothbrush attachment for dental floss and tongue cleaner, an electric toothbrush with a timer to make sure the kids brush for two full minutes.

Toilet tissue dispenser: Consider a tissue dispenser to dispense the required amount of toilet tissue by the press of a button; a toilet tissue dispenser with on-demand application of moisture, cream, perfume, etc.; water conservation in a toilet tank; or odor-reducing ideas for a toilet bowl.

Baby or adult diapers: Every year, diaper designs and their claims are changing; this shows that consumers are dissatisfied with current products or want something better. As a consumer, you can come up with the features you would like to see in a diaper.

Sleeping comfort: Consider pillows and mattresses designed with temperature control, earplugs, eye masks, etc.

Drain clogs: Think of new methods for unclogging drains.

_____
_____
_____

Painting made easy: Consider better designs for paintbrushes and rollers. You could invent a built-in mask or guide on the paintbrush to eliminate time-consuming masking tape application prior to painting.

_____
_____
_____

Ironing clothes: Think of better and easier tools to iron and fold clothes.

_____
_____
_____

Exercise equipment: Think of how to improve exercise equipment like treadmills, bicycles, mini-gyms, etc., to make them safer and make it easier to exercise. Consider ideas to use equipment without getting bored.

_____
_____
_____

Design house items: Invent low-cost sensors and remote controls to operate household appliances.

_____
_____
_____

Indoor plants: Create a mechanism to water indoor plants (more efficiently than what is currently available) while you are away on vacation.

THE INVENTOR IN YOU

Solar power: Solar technology is very advanced. How can you apply this technology to daily indoor and outdoor activities?

Garage: It would be nice to have a garage organizer to make it easy to sort and store tools and other items in the garage.

Indoor pets: Create a better way to handle cat litter or a comfortable way to leave pets in the house while away for short vacation.

Preschool and Grade School

Homework: Think of ways to make kids enjoy doing homework by having fun, mixing homework with games, toys, songs, drawings, etc. Create an easy way for the parents to check if the kids have completed homework correctly—maybe something as simple as a scorecard or, alternatively, an advanced computer program.

Outdoor Applications

Automotive: Sun visors may not always block the sun, especially during morning and evening commutes. This is even worse if you are not a tall person. Simple fixtures like an extendable shade can make your ride more comfortable. Similar fixtures can be designed for side and rear windows. We always look for better spill-proof cup holders. How about a solar-powered vacuum cleaner for the car that gets charged while driving?

_____
_____
_____

Sports equipment: There are plenty of opportunities in this field. Every day, you see new or modified sports equipment. There is a real need and demand for new and different sporting equipment. You can make your own wish list.

_____
_____
_____

Toys: This is an area of unlimited opportunity for inventions. You may want to go to a toy store and spend some quality time studying various toys. There are two ways you can look at this—coming up with new toy ideas based on popular toys or trying to improve existing toys. You may think of combining multiple toy features for your own idea. Let your imagination go wild.

_____
_____
_____

Tools: Every time you use a tool, you hear two types of comments: "This is a great tool, and I love it!" or "This is the worst tool I've ever used. Who invented this?" We have gone from the basic straight-edged screwdriver to multiple heads that can fit on one handle. So much

THE INVENTOR IN YOU

work has been done on this single tool alone, so imagine the endless opportunities for invention in this area!

Lawn maintenance: Many areas need help here—weed control, lawn mowing, fertilizing, an easy way to trim and shape hedges, water conservation, and storing and using rainwater for irrigation.

Pets: Think of easy ways to walk and clean up after dogs.

Safety: You can revisit all the topics that we mentioned so far and look for more opportunities in each item with safety in mind. Safety is an important factor in everything we use. We hear about so many people getting hurt from the things that we use every day because of faulty equipment, poor design, or improper use. Everything consumers use would benefit from foolproof safety features. You probably already know where additional safety features are needed in some of the equipment you use. You can also focus on safety equipment such as helmets, shoes, pads, shin guards, etc.

Recycle and reuse: Find ways to reuse or recycle the items we otherwise throw away.

Travel: Think of efficient ways to pack clothes for vacation, better design for luggage, ways to enjoy long flights, ideas to reduce jet lag, etc.

_____
_____
_____

Snow removal: Create an easy way to remove snow without strenuous shoveling.

_____
_____
_____

Try to expand this list by adding more areas such as swimming, fishing, entertainment, medical, etc.

Generating Ideas and Solutions

By generating a wish list, you probably have identified many opportunities or unmet needs for an invention. Each item in your list is a problem that needs to be solved or an unmet need to be fulfilled. At this point, you need to decide which item on your wish list you want to tackle. With this exercise, you will also have a few possible ideas for a solution. Rather than focusing on multiple ideas, prioritize them according to their significance and pick the top item with a chart such as the one I used in chapter 5 to do that. Then concentrate on the top idea and focus on the details that will lead you to take this idea to fruition or fulfill the unmet need. You will find many useful tools in chapter 6 that will help you generate ideas. Using these tools, you should be able to think like an inventor and generate various solutions.

If you are working in an industry, your chances of inventing are much greater. Many companies encourage all their employees to come up with suggestions for improving their products and operations. I have seen many good suggestions come from maintenance staff, machine operators, and workers on the assembly line because they handle the products and processes every day and know more about them than anybody else. That means they can probably identify weaknesses better than others and may have ideas to correct them.

You don't necessarily have to design a new product to meet your goal. Instead of developing a new product, you may be able to modify an existing product by adding something to it or taking something from it. An example of this is my company's attaching aluminum foil to Bubble Wrap and thus coming up with a product that offered reflective as well as conducive insulation.

You may even be able to combine two or more existing products to satisfy your unmet need. As I mentioned earlier, Remya Jose did this when she hooked up a bicycle to a washing machine; she got around the need for electricity in this manner. Another example might be a vacuum cleaner/dryer combination that could vacuum as well as dry a wet floor. Once you start thinking this way, I am sure you'll come up with many product ideas for new uses and many new ideas for invention.

With a creative mind, you may also come up with an invention without having a specific goal in mind. This happens many times when an experiment has failed or gone wrong and has produced a negative result. Instead of ignoring the failure, you can turn it around and make something useful out of it. An example of this was the Post-it note I wrote about earlier; a failure became a success.

It is also possible for you to create an invention by finding a new application for an existing product that was intended for something else. Bubble Wrap was originally intended to be a type of wallpaper, but I believe someone published a book compiling a thousand uses for Bubble Wrap. How many uses can you list?

One way to create this type of invention is to do a simple exercise. Every time you see a product, make a list of features that product has and functions it is capable of performing. For example, when you look at a vacuum cleaner, you see only that the air is being sucked into the unit; you may not realize the air has to go somewhere after filtering the dirt. If you look at the other end, you will notice a powerful airflow blowing out of the unit. Here is an opportunity for somebody to make use of that feature for something else. If you start to look at any product in a similar way, you are likely to find some additional features not utilized in that product. These are all clues for your next invention.

Here are some helpful hints in generating solutions: Try to understand all the problems before generating solutions. You cannot come up with the right solution if you don't fully understand the problem. Consider all the possible solutions and don't prejudge them. Remember that there are many ways to get from point A to point B, and the shortest route may not necessarily be the most efficient. Have an open mind. Let your imagination go wild. Look at all the possibilities.

Once you have decided on a possible solution to the problem or an idea to fulfill the unmet need, try to get a mental picture of your idea, write it down, and draw pictures. All these steps will help you to finalize the design.

At that point, you may want to make some prototype models of your design to make sure it is viable. If you are lucky, you may come up with multiple ideas for solutions. Once you have a working prototype, you have concluded the invention process. Now you are an inventor.

Depending on how valuable you think your invention is, you can take the following steps. First, find out if this is an original invention or if anybody has previously created a similar invention. Conduct a prior-art search as described in chapter 7. Don't become overconfident; your chances of finding some prior art are great, but you have done well if you don't come across any prior art. Don't be disappointed even if you come across prior art. Following the instructions in chapter 7, you may be able to design around the prior art and possibly get a patent granted for your invention. Even if somebody has come up with similar idea before you did, you are still an inventor as long as you came up with this invention independently without any knowledge of the prior art.

Congratulations on your new invention! You are now in the inventor's league.

www.ingramcontent.com/pod-product-compliance
Lightning Source LLC
Chambersburg PA
CBHW020653220526
45464CB00001B/421